Lost Sounds

Lost Sounds

The Story of Coast Fog Signals

Alan Renton

Whittles Publishing

Typeset by
Whittles Publishing Services

Published by
Whittles Publishing,
Roseleigh House,
Latheronwheel,
Caithness, KW5 6DW,
Scotland, UK

ISBN 1-870325-83-4

Printed by Interprint td., Malta

Contents

Foreword

by Kenneth Sutton-Jones, Hon.Mem., IALA

The sight of a lone building on a headland often interests onlookers. If it were to protect the seafarer from local dangers, it would be painted so as to be conspicuous, bear a light flashing with an identifiable character and once would have been manned by a special breed known as lightkeepers. The thought of their lonely vigil would pluck the heartstrings of most romantics. If this vigil was maintained in an offshore rock lighthouse or aboard a lightvessel, subject to the impact of the sea's motion, it would heighten the mood of the romantic and engender a sense of awe and admiration that men can endure such privation to safeguard the sailor.

Once fog enveloped the scene, an eeriness and sense of foreboding accompanied the gloom which was intensified by a low and weird 'Booomming' sound from the monster no longer visible yet manifestly still there. Such gargantuan blasts came singly or in groups, to identify the location of what the eye was deprived of seeing in the fog or rain.

The 'Booom' could be an explosion, a crescendo eventually reaching full bellow, or cease with a sharply-descending 'grunt'. All gave the the station a means of identifying itself to those anxiously listening aboard vessels at sea that may be running into danger. It seems amazing that lightkeepers and lightsmen could tolerate such ear-shattering noise, a noise which local residents considered to be unwelcome, yet an 'intrusive necessity'.

Through technological advances and sheer economic necessity, the means of creating such signals has changed and with it the romance has been eroded. The lightkeeper's duty is over and he has had to retire, some would say a victim of progress. Although position fixing at sea has been enhanced by the introduction of satellites and ingenious instrumentation, a visual signal is needed as a means of assurance, if not necessity. The flashing display of the lighthouse, light-float, beacon or buoy, although automatic and unmanned, is still a feature but without its lonesome watchman, the romance loses much of its lustre.

The Racon, providing its flashing imprint upon the sailor's radar screen has banished the need for the powerful sound signal. The boom is a rarity except that it now provides a two-mile 'Peeep' of higher pitch automatically at some stations in case a ship collides with it!

Alan Renton, who among his many talents is a specialist in sound recording, visited many of the stations using sound signals before the powerful equipment was dismantled. So — is this subject, so intriguing to many a layman, closed for ever?

This volume describes almost all known sound signalling devices produced over the centuries for use at sea. Not only does Alan Renton cover the sounding instruments – gongs, bells, rockets, explosives, reeds, sirens, typhons, diaphones and electric emitters – the coding units and entire installations, he also describes complete lightstations where sound signals were required. I believe such a work to be unique. It is fascinating and provides a most valuable review of past fog-signalling at sea.

My own involvement in the design and provision of such aids to the mariner over a period exceeding sixty years enables me to realise the value of this book as a record of the past and it should be retained in the libraries of lighthouse authorities worldwide.

Accordingly, it is with great pleasure that I commend this work to all with an interest in lighthouse engineering, specialists in sound propagation and all deprived romantics who know that this eerie 'booomming' sound on a foggy day is, alas, now part of history.

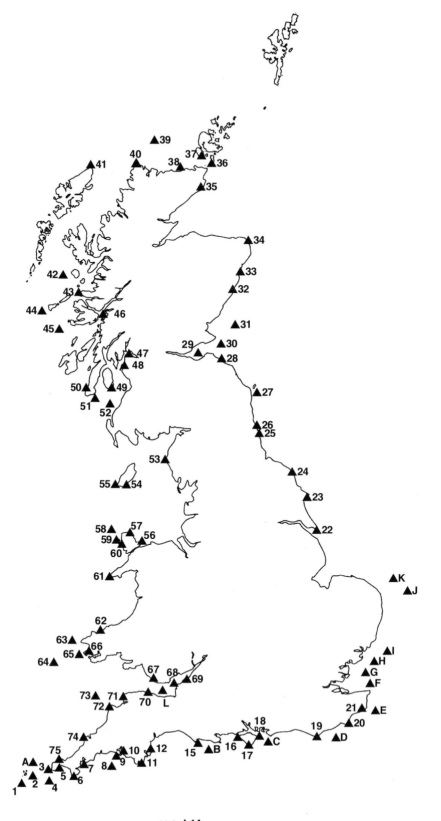

Lighthouses and lightvessels referred to in the text

Introduction

Fog signals around the coastline were once of great importance but the sounds that helped mariners, otherwise lost in the fog, have now been superseded by improved aids to navigation and a unique aspect of maritime industrial archaeology has passed into history.

This book is the story of how fog signals came about and how they worked, why sound signals were necessary and why they were ultimately superseded.

By the early 1990s the programme to automate all British lighthouses was well advanced. Various types of fog signal machinery were a casualty of automation and before the surviving different types were all lost, sound recordings of them in operation needed to be made. The Lighthouse authorities and the British Library National Sound Archive agreed and, in 1993, I was awarded a Leverhulme Research Fellowship to enable the recording project to be carried out. The fog signal recordings are now held by the National Sound Archive and readers curious to know more can go and hear the wonderful sounds for themselves!

The book finally resulted from the research started whilst making the recordings. As the manuscript was largely completed before the last lighthouses were automated in 1998, the book describes the operation of a fog signal at a manned lighthouse in the last years of service – snapshots of the machinery and keepers before both became redundant. The huge old air trumpets are gone but the fog signal is not entirely extinct – a number of automatic electric sound emitters still form part of the network of navigation aids provided to the mariner by the Lighthouse authorities.

Lost Sounds is a lighthouse tour of Britain – each chapter concentrates on the different fog signals once used at a particular site to tell the developing story of new design and implementation.

In the 19th century, British influence was widespread through trade, colonial power and industrial technology but not all fog signal inventions and improvements were made by Victorian engineers. In the early days of the fog signal, the French and American lighthouse authorities were ahead of the British lighthouse authorities in some respects and developments in Canada, Germany and Sweden made significant contributions to the story of fog signals early in the 20th century. The story of fog signals on the British coast reflects the bigger picture of the story of fog signals worldwide.

My lighthouse tour took me to many sites, ancient and modern, manned and automatic and many attendants and keepers gave me their time and cups of tea in the course of my research. I would like to thank them all for their friendly assistance and I would also like to thank all those other current and former staff of Trinity House, the Northern Lighthouse Board, the various harbour authorities and private companies, libraries, museums and archives who have helped me with the project. I am especially grateful to Jane Wilson, Colin Wortley, Gerry Douglas-Sherwood, Ian Peattie, Kenneth Sutton-Jones, the Leverhulme Trust and my long suffering family!

Alan Renton

Chapter 1

Trywn Du

and the fog bell

The warm front passes over and the mild air brings in low cloud behind the showers. As the warm humid air moves over the cool sea surface, the air temperature falls below its dew point and fog appears. From the boat there is nothing to see except varying densities of blurred greys. From the aeroplane the fog looks like clouds which have settled on the sea in dense heaps, but in places there are thin patches enabling glimpses of the water. The exhaust from the boat has left a trail in the thin fog like the vapour trail of an aircraft.

Looking out to sea from the land the grey water merges with the grey sky, though there is some slight movement of the light and dark shades. Nearer the land there are glimpses of cloud tops and dim patches of hazy blue sky appear. From the lighthouse it is obvious that the fog is slowly clearing and the visibility is improving. The fog marks[1] up and down the coast are now visible and the stray wispy patches of fog which were blowing in have all gone. The fog signal has stopped.

The lighthouse stands impressively during the day and even more spectacularly at night. By comparison, the fog signal is unassuming and once the invisible sound it produces has stopped there are few visual signs of its presence to the casual observer. At some lighthouses there are bits and pieces poking out of the buildings, some of which resemble the mouths of huge trumpets though the function of other objects is less apparent.

Of the many hundreds of sound signals once used during poor visibility around the coast at major lighthouses, there are now only a few dozen. Powerful compressed-air fog signals operated by large engines have all been replaced by compact automatic electric emitters, where a fog signal has been retained at all.

Bells are believed to have provided the earliest fog warnings but surviving fog

---

[1] Prominent marks such as rocks or headlands at a known distance from a lighthouse were used as references to help the keeper determine when to start and stop fog signals.

Trywn Du Lighthouse and fog bell.

bells are very rare, with only a few remaining at harbour lighthouses. The last fog bell in service with Trinity House is at Trywn Du lighthouse on Anglesey.

The early use of bells as navigational aids remains speculative but bells have long been used aboard ships and it is reasonable to believe that they were similarly used near harbour mouths. In the calm conditions that often accompany fog, all kinds of sounds reaching a boat from the land might have been useful in trying to establish an approximate position and set a suitable course. If church bells rang in the harbour, local seamen could sail towards the sound and under favourable conditions the bells might be audible for a few miles out to sea.

One of the earliest established dates for a fog bell is confirmed by drawings from 1766 which show a bell in a purpose-built wooden tower which was constructed at Nidingen in the Baltic.[2]

From 1777 on the Northumberland coast, a bell was regularly sounded from the south turret at Bamburgh Castle during fog, with the approval of Trinity House in Newcastle.

The Commissioners of Northern Lighthouses issued a Notice to Mariners on 1st February 1811 in relation to Robert Stevenson's Bell Rock Lighthouse:

---

[2] Christie, R. and Hague, D.B., *Lighthouses: their architecture, history and archaeology*, Gomer Press, 1975.

> *'... during continuance of foggy weather and showers of snow, a bell will be tolled by machinery, night and day, at intervals of half a minute'.*

By 1820 several bells had been installed at American and European lighthouses and the pile lighthouse on the Maplin Sands was fitted with a bell in 1838. From 1841 the installation of bells was approved by Trinity House for lighthouses under its administration. Fifteen lighthouses on the English and Welsh coast were fitted with fog bells by 1867 including Casquets, Plymouth, Bishop Rock and other lights in the Thames approaches.[3]

A fog bell was installed at Trywn Du in 1867 at a cost of £146 5s. 0d. and a reference to it first appeared in the Admiralty List of Lights in 1871:

> *'A large bell will be sounded in foggy weather; three times in quick succession every 15 seconds'.*

The lighthouse was first proposed in 1748 and was to be built on Puffin Island off the north-east coast of Anglesey. However, it was not until after the 1762 Liverpool Dock Act that the Pilot Committee seriously looked at leasing the island[4] both to erect a lighthouse and to establish a depot and rendezvous point for vessels approaching Liverpool. The Mersey Dock Board was responsible for developing many of the lights along the north-west coast of Wales and had pilot vessels patrolling these waters, including one off Puffin Island.

Trinity House finally consolidated proposals in 1833 and on June 28th 1838 'Tywn Du' (*sic*) was first shown. A fixed red light warned of a rock ledge between the lighthouse and the island in the narrow channel leading into the Menai Straits. The round tower was built between 1835 and 1838 to James Walker's designs and has vertical walls and a castellated parapet on the gallery. The fog bell was later suspended from the gallery.

In 1863 and 1864 Trinity House carried out a series of fog bell trials following similar ones undertaken by the French lighthouse authority. Observers were obliged to note any improvements in the loudness of a bell and the results were noted to detect the effectiveness of the different sizes of bell, whether the use of hemispherical reflectors improved the signal and if different weights of hammers falling from different heights on to the sound bow of the bell produced better results.[5]

Close attention was paid to the prevailing wind speed and direction during the trials and a range of figures was compiled. The results confirmed the observations

---

[3] The figure excludes the number of bells at harbour lighthouses. See *Return of all fog signals now in operation under the General Lighthouse Authorities of the United Kingdom*. House of Commons, 21st March 1873.

[4] The Liverpool Dock Act of 1762 granted powers to purchase land in Lancashire, Cheshire and Wales and discussions even considered the Calf of Man.

[5] Steel bells were also tested but found to be inferior in all respects bar cost. Some steel bells were, nevertheless, used and they were more widely used at American lighthouses, despite their vulnerability to corrosion.

made by the French: the audible range of bells could be improved by increasing the number of strokes[6] and by using a reflector to help concentrate the sound, but their range remained limited. It was generally accepted that striking the bell with a hammer was more efficient than swinging the bell. The maximum range of a 10-hundredweight bell was less than a mile over an intervening fresh breeze.

The mixed results did not detract from the developing interest in installing audible warning signals to assist shipping when poor visibility obscured the lighthouses, and bells seemed to be the most suitable type of signal. As an interesting coincidence, at about the same time there was a revival taking place in bell-making in England, supported by the building of new churches and the restoration of old ones along with the construction of impressive Victorian town halls.

Of the many bell foundries in business in the 1850s and 1860s only two have survived.[7] Bell casting and tuning remains a very specialized skill. The smallest fog bells cast were about 3 cwt but the average size was 7 cwt. The bell at Stevenson's Bell Rock in 1811 was 5 cwt.

Reports on the usefulness of the early bells in service were evaluated by the lighthouse authorities, but these reports were inconclusive. An Irish bell of two and a quarter tons with a sixty-pound hammer which fell ten inches was audible only one mile to windward against a light breeze, although another report from a Holyhead steam packet claimed to have heard a similar bell at over three miles above the sound of the vessel's paddle wheels.

The Secretary to the Commissioners of Northern Lighthouses, Alexander Cunningham, believed that there was no practical alternative to the bell, particularly

> Bell metal consisted of 77% copper with the rest being made up of tin, some zinc and a certain percentage of metal from old bells. If the tin content was too high the bell metal could be too brittle and the crack in Big Ben in the tower of the Houses of Parliament in London, which was cast in 1858, is attributed to this condition. Once cast, a bell was tuned by paring away concentric rings of metal from the inside of the sound bow and the waist by using a specialist vertical lathe. The effect of this process on the sound produced by the bell was repeatedly measured with tuning forks.
>
> The characteristic sound of a bell is composed of a range of distinctive harmonics. A nominal tone initially dominates when the bell is first struck but this quickly dies away. The 'hum' note which remains is made up of thirds, octaves and other partials of the fundamental note and the nominal. The bell tuner refines the relationship of the greatest number of harmonic tones with the fundamental.

[6]   The French trials undertaken by the Corps des Ponts et Chaussées in 1861, established that sixty strokes per minute increased the effective range of a bell by nearly 30%.

[7]   Taylors of Loughborough and the Whitechapel Bell Foundry in London.

at rock lighthouses such as Skerryvore and Bell Rock. A fog bell was, therefore, better than nothing although he was doubtful that:

> '... *either bell has been the means of saving a single vessel from shipwreck during fog.*'[8]

On rock lighthouses the bells weighed between 5 and 7 cwt and were suspended from the lantern gallery about 100 feet above sea level. One of the criticisms concerning the effectiveness of bells at rock lighthouses was down to the positioning of the bell on the gallery on the least exposed side of the tower. The sound of the bell was screened by the tower in the very direction it was most likely to be needed.

At exposed rocks, green water not uncommonly reached the height of the lantern, and bells were placed on the lantern gallery opposite the prevailing sea as a precaution against accident. In January 1860, the original 3 cwt bell at the Bishop Rock was '... wrenched off by a sea striking it ...'[9] during a storm and the 1863 List of Lights stated:

> '*The Bell, formerly sounded in foggy weather, has been washed away, and is not replaced*'.

A larger bell replaced the original in 1864.

A bell was proposed for the Eddystone Lighthouse in 1865, although it was 1872 before a 5-cwt bell was installed at a cost of £327 5s. 10d. When Douglass' new lighthouse was nearing completion at Eddystone in October 1881 two huge bells were installed on opposite sides of the gallery. By 1891, however, the bells were used only to supplement the new explosive fog signal in an emergency.

Bells at other major rock lights remained in use until replaced by the more satisfactory explosive fog signal. Bishop Rock had an explosive signal in 1887 but Longships lighthouse waited a further ten years for similar equipment. Wolf Rock retained its bell until 1904, when a reed fog signal was installed to replace the clockwork mechanism and the 7-cwt bell. The bell sounded three strokes every fifteen seconds.[10]

The idea of giving each station bell a distinctive character was adopted in 1871 and existing striking mechanisms were modified. The new character assisted the mariner in distinguishing one signal from another and helped identify specific lighthouses. The character consisted of a particular number of strokes within a given period and this information was published in the Admiralty List of Lights.

---

[8] Cunningham, Alexander, *Fog Signals*, paper presented to the Royal Scottish Society of Arts in 1863.

[9] *Engineering*, 9th January 1874.

[10] The Wolf Rock was '...struck by two hammers worked by machinery fixed in the pedestal of the illuminating apparatus but independent of that for rotating the latter. For the purpose of giving this signal a distinctive character for the station, the mechanism is arranged for striking the bell three blows in quick succession at intervals of fifteen seconds'. Douglass, James, *Minutes of the Proceedings of the Institution of Civil Engineers*, paper no. 1268.

Fog bells at the new Eddystone Lighthouse in the early 1880s.

At stations where the character was to be two or three strokes in a given interval, the hammer was positioned on a central shaft and had shaped heads to allow even striking of the curved surface of the bell. A clockwork drive turned a cam shaft which lifted the hammers in turn before letting them fall under their own weight against the sound bow of the bell. It was established that hammers striking the outside of the bell proved more effective than a clapper inside. Once the hammer had struck the bell it needed to be lifted clear and a flat leaf spring was placed between the hammer and the bell. The spring held the hammer clear of the bell until it was raised by the cam for the next stroke.

Clockwork striking machinery usually had separate weights but it was possible for the optic drive machinery to sound the bell as was the case with the first fog bells installed at Bell Rock.

The bell at St Anthony Head Lighthouse, circa 1890. Courtesy Royal Institute of Cornwall.

The elegant Georgian building at St Anthony Head was originally called Falmouth Lighthouse and the tapering octagonal tower, completed in 1835, was designed by James Walker. The 1865 bell was activated by a clockwork release mechanism beneath the revolving lamps and reflectors and the weights ran down a 20-foot shaft in the tower wall. In 1882 a two-ton bell replaced the original one but retained the same character of four strokes every alternate half minute. The original weights and clockwork may also have been retained. An undated Trinity House engineering drawing shows the shaft for the weights and the large bell. Another drawing from 1896, signed by Thomas Matthews, the Engineer in Chief to Trinity House at the time, shows new actuating machinery to operate both the revolving apparatus for the lantern and sound the fog bell.[11]

11   The Matthews drawing shows the character of the revolving light to have been one flash every twenty seconds and the fog bell was four quick strokes every minute. To drive the optic, weights totalling 7 cwt fell 38 feet. By engaging another drive wheel and larger weight (15½ cwt) the machinery could drive the optic and the fog bell. The 7 cwt weights required 170 revolutions of the winding handle but the 15½ cwt weights required 395 revolutions to wind them up the 38-foot drop! *Trinity House Drawing* no. 7456, 9th June 1896.

When a new static lens later replaced the revolving apparatus, the height of the lantern was raised and a drum ventilator replaced the earlier ball finial cowl on the conical roof. The actuating machinery below the lantern was replaced or modified at the same time as the new optic lamp had an occulting character produced by a clockwork eclipsing mechanism.

The large bell had a diameter of five feet and was secured to a large iron girder on the balcony. There is no longer any trace of the girder or the securing bolts, but the stone work on the gallery shows the position of the bell with evidence of the entry port for the chain which linked the machinery and the hammer. Beneath the optic are a few bolt holes in the floor where the machinery stood.

When the bell was removed in October 1954 and the experimental electric fog signal was installed, Trinity House agreed to make a gift of the bell to Penwerris Church in Falmouth. Unfortunately, after transporting the bell to the church it was not practical to install it and it was subsequently dispatched to Taylor's foundry at Loughborough and melted down.

A striking mechanism that was capable of producing a distinctive character was demonstrated in the United States by Mr D Johns, who took out a patent as early as 1849. There were other patents for similar equipment and Stevens and Daboll introduced their own clockwork machinery for striking bells. Stevens made church and town hall clocks and Daboll subsequently made major contributions to the development of fog signals.

Unlike the situation in Britain, bells were used at many shore stations in America. Some American fog bells built in the 1850s and 1860s,[12] had separate bell houses of extraordinary design, some looking like very tall pyramids, which housed the bell at the top and had the weights running down inside. Other buildings had a tower for the weights but the bell house was on ground level with the bell mounted clear of the end of the building. Inside the bell house was the striking machinery and a panel on the wall gave access to the surface of the bell located outside. When fog set in the hammer would swing through the aperture to strike the bell.

Instead of building a tower, the bell house at some stations was positioned near the cliff edge so the weights could be run straight down the cliff. This arrangement was used at Alcatraz but salt water rusted the wires and weights.

In 1862, at Start Point lighthouse in Devon, a bell house was built near the cliff edge and the weights for the clockwork ran down the high cliffs. This bell was one of the largest used by Trinity House and the clockwork mechanism struck the bell 48 times every minute. It must have been some job to walk from the lighthouse along the cliff path in poor weather to the bell house to wind up the large weights. A siren replaced the bell at Start Point in 1877 but the bell house still stands. It can

---

[12] Bells continued to be installed at many stations despite all their drawbacks. By 1900 there were 219 fog bells in service in the USA.

no longer be reached, however, as a landslip in 1989 demolished part of the walled pathway leading to it, besides destroying the siren and fog signal building.

In 1879 the 32-cwt bell was removed from Start Point and transported down the coast to Plymouth where it was installed in the Breakwater Lighthouse. The bell was valued at £192 and new clockwork machinery was made at a cost of £95 with a further £90 spent on fittings. The original 7-cwt Breakwater Bell from 1845 was apparently extremely useful in fog – local pilots could navigate by it.[13] This bell was one of the first to be modified after 1871 to produce the distinctive characteristic variations. It was removed to the Gunfleet lighthouse when replaced by the bigger bell.

Ownership of the lighthouse on Plymouth Breakwater was passed from Trinity House to the Naval Dockyard in 1993 and the bell at the Breakwater was still in use until 1994 when it was replaced by an electric emitter. The Breakwater Light was automated many years ago and was later monitored from across the water at Penlee Fog Signal Station. Acetylene gas was used from 1920 as the illuminant. Prior to the recent modernisation the last major work on the automatic lighthouse was undertaken in the 1960s, when the bell was modified.

Some clockwork-operated fog bells were converted to gas operation to function automatically. In the entrance room of the Breakwater Lighthouse, carbon dioxide gas bottles were installed and connected to a striker mechanism below the bell via a controlling flasher mechanism similar to those used in acetylene lanterns on buoys. The flasher regulated the gas supply to release the striker and a piston was fired by the expanding carbon dioxide against the inside of the sound bow once every fifteen seconds.[14] A similar system was used on light floats and some bell buoys also used carbon dioxide from the 1930s and 1940s.

As early as 1901 the Northern Lighthouse Board had installed an unattended light-float off the west coast of Scotland at Otter Rock. Otter Rock light-float was positioned at the south east of Islay and was the first such vessel in British waters. Pintsch's gas system was used for the light and the 6-cwt bell was fitted in a frame on the deck from which four large clappers were suspended. The rolling of the vessel caused the clappers to strike the bell. Wave-actuated bells are still in service on some buoys although many of the 3-cwt bells have been removed. Redundant bells at the Trinity House depot at Swansea are characteristically worn where the wave-actuated hammers have mercilessly recorded the violence of the sea. Of course, in a calm sea the wave-actuated bell was ineffective and David Stevenson resolved this problem on the Otter Rock light-float by using a gas system to operate the striker.

---

[13]   The Breakwater fog bell '...is constantly rung in foggy weather; and such is the experience of the pilots, that as according as the Sound of the bell reaches them they know at once in what part of the sea they are'. Esquiros, Alphonse, *Cornwall and its Coast*, 1865.

[14]   A fog detector at Penlee Point started the Breakwater fog bell automatically from the 1960s.

> On its passage to the lantern from the storage bottles, the gas was directed via a spring-loaded control valve. The build up of gas pressure in the valve chamber forced an elastic diaphragm upwards to move a lever which closed the inlet valve and opened the outlet. The pressure release tripped a spring to fire the hammer against the bell. At the same time, the diaphragm was depressed by a spring and the inlet valve was opened to repeat the cycle. The bell at Otter Rock was struck three times every minute.

In British waters major buoys have all been converted to solar power but the AGA flasher was used on hundreds of buoys formerly lit by acetylene and it was also used at the acetylene lighthouses, including Trywn Du, until recent times.[15]

AGA recommended their flashers were inspected every ten years but there are examples of the flashers working reliably for fifty years without attention! The Swedish AGA company was created by Gustav Dalén, who developed the use of acetylene for lighting.[16] The Dalén flasher came into general use from about 1906[17] and Dalén also invented the sun valve which automatically opened the gas supply to the burner at sunset and closed off the gas at dawn.[18]

The use of stored gas at lighthouses liberated the invention of many engineers. Since a pilot light had to be kept burning there was a constant flow of gas, however small, and delightful engineering used this energy through flasher systems, pumping systems and gearing to provide surprising examples of automatic operation. The gas could be made to turn the optic, wind the clockwork weights and could even sound a fog signal.[19]

---

[15] The following Trinity House lighthouses were lit by acetylene: Berry Head, Caldy, East Usk, Egypt Point, Farne, Godrevy, Hurst, Penninis Head, Plymouth Breakwater, St Tudwalls, Trywn Du.

[16] At the time, Scandinavia was the world's largest producer of calcium carbide from which acetylene is produced. Acetylene gas cost three farthings per cubic foot in 1910.

[17] The AGA flashers varied in size but would typically be about five inches in diameter. Two chambers operated to regulate gas pressure and deliver an intermittent flow of gas. The diaphragm in the flasher was made of kid skin saturated in paraffin wax, on a thin steel plate. Expanding gas forced the diaphragm upwards. The diaphragm had a steel pin connecting it to a lever inside the chamber which had a magnetized tip. Precisely engineered input and output valves at the extent of the travel of the lever guaranteed a precise switching between open and closed. The lower chamber operated in a similar fashion to regulate the fluctuation in gas pressure from the supply bottles which stored gas at high pressure. If combinations of these flasher units were used, the gas output could turn cams or produce complex characteristics for a flashing light or fog signal.

[18] The sun valve used differential rates of expansion of five copper rods to open and close the gas supply. The four outer rods were gold plated and supported a black inner rod connected to the valve. The black rod more readily absorbed heat than the polished rods, and daylight was sufficient to make it expand downwards and close the valve to cut the gas supply. As night fell the rod contracted and the valve opened. These instruments were very sensitive and extremely reliable – only with the discontinuation of the use of gas did they become redundant.

[19] Matthews used gas to rotate an optic at East Usk in 1893. Pressurised gas escaped into one of two cylinders which drove pistons up and down with valve gear to open and close the supply. The piston rods were attached to a rocker beam which wound up clockwork to turn the optic!

Dalén perfected the safe storage and transport of acetylene which is stable in acetone, but acetylene could also be produced on site by the action of immersing carbide in water. Some early automatic buoys produced their own acetylene gas and similar plant was installed at several lighthouses.[20] The pioneering use of acetylene enabled the automation of many lights early this century, particularly in Scotland.

There was another device in use at a few stations at this time, the hydroscope. This device was sensitive to moisture in the air and could be used to start the fog signal. The automation programme may have accelerated in recent years but it was started a long time ago.

Trywn Du was automated in 1922, though an attendant lived in the keepers' cottages on the shore nearby. A new automatically-operated fog bell was installed in 1921. The wave-washed lighthouse is only a short walk across the weed-covered rocks at low tide. The stepped base of the tower[21] has alternating half steps which are bordered by bronze dog steps and it is a steep climb up into the entrance room.

On my visit the entrance room still contained the acetylene bottles for the lantern and the carbon dioxide bottles for the fog bell, but the lighthouse has subsequently been converted to solar power and the gas bottles were last used in 1994. Since May 1995, the lighthouse has been remotely monitored from the Trinity House Operation Control Centre at Harwich.

Most visits to the lighthouse were enabled by walking out to the tower but a boat was used to deliver the heavy gas bottles which were manhandled into the entrance room after being winched up to the door by the hoist protruding from the third floor. Within the small entrance is an inner room which formerly contained nine carbon dioxide bottles on line to the fog bell, with a further nine disconnected, depleted bottles arranged around the wall. The acetylene burner in the lantern room was connected to four bottles and operated automatically by the sun valve positioned on the lantern roof. The illuminant was a three-cluster open-flame burner[22] and the lighthouse had a fixed first order catadioptric lens.

The automated electric lighthouses are heated to help prevent deterioration and damp but the acetylene lighthouses, like Trywn Du, had no heating and the cold dark interior felt in keeping with its battlements above. The lighthouse was sparsely furnished after the keepers were taken off, but the kitchen still retains the original cast iron range and the curved Welsh dresser. The building shares the elegance of Walker's St Anthony Head lighthouse.

---

[20] Farne Lighthouse produced its own acetylene from 1910 and, still visible on the cliffs (not to be confused with guano!), are streaks of the white sludge by-product which was thrown over the side. The carbide sludge was used to paint lighthouse walls. Similar equipment was used at Hurst Point.

[21] Walker devised the stepped base to break up the strength of waves crashing against the tower, and the design was later used at Wolf Rock and other rock lights.

[22] By crossing the burning jets, a richer air mixture helped to give a brighter, less yellow light. The cluster burner was also used at Berry Head, but other acetylene lighthouses used small silk mantles.

The bell and striker mechanism at Trwyn Du.

On the third floor is the hoist and above this is the keepers' bedroom. On my visit the bedroom was also empty of most furnishings, unlike other stations where the accommodation actively awaits the next maintenance visit and still looks 'lived in'. Trwyn Du felt like a preserved historic monument. The furnishings may always have been sparse and simple. The bedroom has a small fireplace with bunks and a crude ladder to the floor. The doors to the rooms are all curved to the radius of the inner walls.

The service room, immediately below the lantern, has a square sash window which overlooks the channel and Puffin Island. Carved in the stone lintel above the window is the same verse as at Smeaton's Eddystone:

'*Except the Lord build the house they labour in vain that build it. Psalm CXXVII.*'[23]

On my visit, this room contained the steel pipework linking the gas bottles with the burner in the lantern above, and the AGA flashers which regulated the character of the light and the bell. The iron lantern has vertical glazing bars and after

[23] Smeaton used the same verse inside his 1759 Eddystone Tower. Douglass used the verse in the 1882 Eddystone and it appears later at Strumble Head (1909).

struggling out on to the battlements through the lantern door, there, at last, was the fog bell. There were actually two bells at Trywn Du, the service bell and the standby for use in the event of problems with the main bell. The 3-cwt standby bell had 'Taylor' inscribed on it and came from a bell buoy,[24] along with the cannon striking mechanism.

The main bell weighed half a ton and the striking mechanism unusually sat above the crown of the bell so that the clapper dropped down inside. The clapper was pivoted on a rocker which rested on top of a piston. When the piston fired up at the rocker, the clapper was thrown against the inside of the sound bow. The piston was fired by carbon dioxide released by an AGA flasher.

The accuracy of the flasher became apparent when standing on the gallery. The bell was struck every thirty seconds, day and night, rain or shine, and though bells may be feeble fog signals, the intensity of the sound at close quarters was sufficient to encourage covering the ears prior to the stroke! Within a few minutes of being on the gallery the regular rhythm of the bell became as vital as a pulse and each stroke could be accurately anticipated. Every thirty seconds throughout the duration of the visit the tower resounded.

From inside the tower, the bell sounded a dull thud which quickly decayed, although the faint ring in the air outside could be heard. The round stone tower and the iron and glass lantern made a peculiar resonating chamber. At those remaining lighthouses with mechanically-operated fog bells the sound reverberating inside the building includes the grinding of the mechanism as it lifts the hammers.

There are only a few surviving bells at harbour lighthouses and port entrances. The Port of Tyne Authority has a bell in the stone tower on the end of the South Pier and there is another bell situated on the impressive pile lighthouse on the head of the Herd Groyne.[25]

The harbour lighthouses at Avonmouth and Fishguard have bells which were installed in 1907 by the major lighthouse engineers, Chance Brothers.[26] Electric motors turn the striking mechanisms to wind a chain taut and then trip it to release the hammer.

From their earliest introduction the effectiveness of the bells was questioned by the very engineers who installed them and the inadequacy of a fog bell has been endlessly repeated ever since. It may seem a little surprising, therefore, that the bells were not all replaced by superior signals generations ago – still some survive! Perhaps we have an affinity with bells which transcends mere utilitarian practice.[27]

[24]   Taylors continued to provide bells for Trinity House buoys into the 1970s.

[25]   Herd Groyne strikes once every five seconds and the mechanical striker has two hammers to make for even running of the cam shaft.

[26]   Chance Brothers cast bells at their Smethwick Works but later subcontracted bell making, primarily to Gillet & Johnston at their Croydon foundry, from about 1920.

[27]   When early electric fog signals were being developed in the 1930s they were often made to sound like bells!

Fishguard Harbour Lighthouse and Bell.

Chapter 2

Dungeness

and the early fog horns

Dungeness is a peculiar place with its own distinctive landscape. Travelling to Dungeness on the miniature railway[1] encourages an unusual perception of the countryside and transforms the environment into the Prairies or the Siberian Plains (the train is full of Russian schoolchildren on this visit!). From the drained marshes and past small wooden houses and Japanese gardens the train approaches the old disused lighthouse.

The retreating sea has piled up grey shingle through which a sparse vegetation pokes. In the wind the shingle moves and clicks as little piles of stones collapse. Indigenous birds and insects make unfamiliar sounds, but there is also another sound, a faint hum. The architecture of the nuclear power station adds to the peculiar fascination of the place, underlined by the eerie chord of the electric fog signal – three blasts reverberate off the stones and the power station.

The 'new' lighthouse is near the water's edge, though the beach and foreshore are indistinct and it looks as though the tide might cover the whole area. In fact, Dungeness continues to grow into the English Channel. Rows of blue-green sea kale have colonized the undulating shingle and suggest tide marks left by the retreating sea. The faint hum is overwhelmed by three more blasts from the fog signal.

The sound is produced by sixty 100-watt loudspeakers built into the top of the lighthouse. Three pure tones of 388, 553 and 680 hertz (Hz, cycles per second) are simultaneously sounded to produce the eerie chord. This type of signal was developed by Trinity House in the 1950s[2] and prior to the construction of the new lighthouse there was an earlier prototype here. The lighthouse came into service in

---

[1] Romney, Hythe and Dymchurch Railway.

[2] The development of the triple frequency fog signal is covered in Chapter 13.

Dungeness Lighthouse with integral fog signal speakers.

1960 and it is one of the last lighthouses to have been built in Britain. The 130-foot lighthouse is formed of precast concrete rings, 12 feet in diameter, which are held by tensioned steel cables running through the tower.

The building retains echoes of its predecessors even though stone, wood, iron and brass have been replaced by concrete, steel and plastic. The green, white and black colour scheme is familiar to other Trinity House lighthouses.

Instead of a huge optic, there is a small sealed-beam unit in the lantern room and below the lantern are the 60 fog signal speakers which form part of the tower wall. In the base of the tower is the necessary control and monitoring equipment along with an emergency generator and the fog signal motors. The lighthouse was designed to be fully automatic and in poor visibility, the fog detector starts one of the two alternators which stimulate the fog signal speakers.

The earliest lighthouse at Dungeness is believed to have been a wooden structure with a coal brazier dating from the early 1600s. A brick structure succeeded the first in 1635 and was positioned nearer to the coast. This building was pulled down on completion of a new tower in 1792, which was designed by Samuel Wyatt at about the same time he was building Trinity House at Tower Hill. This huge tower was one of the last ten private lighthouses. An 1836 Act of Parliament abolished privately-owned or leased lighthouses and empowered Trinity House to administer them. The Coke family, which had held the patent for over one hundred and fifty years was paid nearly £21,000 for Dungeness.

When Robert Stevenson visited Wyatt's lighthouse in 1813[3] he was alarmed by the large cracks in the tower caused by a lightning strike. The lighthouse survived, however, and saw the arrival of electric light in 1862.

The 'old' tower, still standing today, was constructed in 1904 when Wyatt's tower was demolished. Shingle deposits left by the sea had resulted in Wyatt's tower being nearly 400 feet from the water by that time. A supplementary light was shown from a small tower nearer the water which was built on top of the fog signal building erected in 1875. The low light and the fog signal building were demolished in 1959.

Dungeness was the site of many Trinity House experiments in the development of navigational aids. The equipment tested here included different lenses and burners, electricity and carbon arc lights and many types of fog signal including bells, guns, explosives, sirens, diaphones and the triple frequency electric foghorn. The first reed fog horn in Britain was installed at Dungeness in 1862.

By 1860 there were only a few fog signals in Europe and America and the majority were bells. Apart from the French trials in 1854 and 1861 and the more recent Trinity House trials with fog bells, there had been very little experimentation by the lighthouse authorities.

---

[3]   Robert Stevenson was Engineer to the Board of the Commissioners of Northern Lighthouses from 1799 to 1843. In his professional capacity he visited many lighthouses around the British coast and visited English and Welsh lighthouses in separate tours in 1801, 1813 and 1818.

Experiments were anticipated in America under the auspice of Professor Henry from the Smithsonian Institute in Washington, which would compare bells and guns and the few steam whistles that were then being used, but the development of other types of signal was at an early stage. Prior to 1870, the only published records of experiments were those at Dungeness in 1864 with guns.

Eighteen-pound guns using three pounds of powder were fired at Dungeness and their effectiveness recorded by observers at various distances from the lighthouse. Four miles away the gun was noted as being 'faint' in fine, clear weather, but when this was compared with a similar gun fired under similar meteorological conditions, the report was distinctly heard at seven miles. Repeated comparisons yielded no more consistency but indicated the unpredictable nature of sound transmission through air under apparently similar conditions, an issue which was to become a recurring problem in future trials and in general service.

The inconsistency in the audibility of sounds under comparable conditions had been previously observed by the Rev. William Derham. Derham submitted various papers to the Royal Society in the late seventeenth century based on his meterological observations and in a paper written in 1708[4] he considered sound transmission in the air and speculated upon the properties of sound transmission in fog.

There was no substantial elaboration on Derham's work for nearly 150 years, until the development of fog signals promoted other Fellows of the Royal Society and eminent thinkers to consider the nature of fog and how it might affect sound.

Derham believed that rain and fog absorbed sound and this theory survived unchallenged. New theories were constructed to demonstrate how sound energy would be lost, either by absorption in the fine suspended water droplets or by endless reflection of the sound off the myriad water droplet surfaces. Dr John Tyndall FRS, was foremost amongst the scientists investigating these theories, and following his lectures to the Royal Institution his book, *Lectures on Sound*, was published in 1867.

Tyndall had succeeded Michael Faraday to the Chair of Natural Philosophy at the Royal Institution and in 1866 he also took over from Faraday as the Scientific Advisor to Trinity House. Tyndall's research built on many experiments he had carried out in his laboratory at the Royal Institution, which suggested analogies between theories on the transmission of light and transmission of sound. He discovered that the velocity of sound was inversely proportional to the square root of the density of the air, and, '… directly proportional to the square root of the elasticity of the air'. He defined elasticity as the force which resisted compression and produced expansion. Heat is generated when air is compressed and Tyndall observed the opposite effect in experiments carried out in 1859. When air expands and becomes rarefied its temperature falls.

---

[4] Derham, W., Experimenta et Observationes de Soni Motu (Experiments and observations on the Motions of Sound), *Philosophical Transactions of the Royal Society*.

Tyndall subjected different gases to thermal radiation in the laboratory to test any changes in the transparency or opacity of the medium to light, and other experiments sought comparisons with the properties of sound transmission in air of variable density.

In his book Tyndall observed that thermal radiation from different surfaces such as grass and bare rock previously heated by the sun, would produce rising columns of air of different density. He thought that the surface boundaries of the different density columns of air would cause refraction and partial reflection of sound. Laboratory experiments[5] established that sound was reflected from the so called 'limiting surfaces' of dense and rarefied air but outside the laboratory Tyndall observed that:

> '...sounds are also reflected from the clouds. When the sky is clear, the report of a cannon on an open plain is short and sharp, while a cloud is sufficient to produce an echo like the rolling of distant thunder. A feeble echo also occurs when sound passes from one mass of air to another of different density'.

The concept of the 'acoustic cloud' was rejected by later scientists, but Tyndall was responsible for identifying many significant areas for further research into the reflection and refraction of sound energy in air masses of different density and humidity, the effect of wind on sound transmission[6] and the possible causes of attenuation and aerial echoes.

Research into the transmission of sound in foggy air was almost non-existent at this time but the British Association for the Advancement of Science wrote to the President of the Board of Trade in 1863, seeking to establish a series of experiments with fog signals. Various opinions were expressed in the report submitted to the Board of Trade, which even included speculation on the nature of the most effective kind of signal – should the sound be steady, or sharp and sudden, or would it be better to have a 'rattling sound'? The authors believed that the effective range of a fog signal should be greater than the distance required for the largest steamer to turn round – at least two miles. It is interesting to note that the report also suggested using underwater signals as it was known that sound travels faster and further in water than in air. The Board of Trade did not act upon the report.

In Britain at this time, the number of fog signals operated by the lighthouse authorities was sixteen, virtually all of which were bells.

As early as 1737 Jonathan Hulls had patented a paddle steamer which was powered by an elemental Newcomen engine.[7] By the end of the eighteenth century steam packets were being built on the Clyde with 3-horse-power engines and sails

---

[5]   Later revisions of Tyndall's book on sound incorporated evidence obtained during experiments for Trinity House.

[6]   Tyndall developed research on the increasing speed of winds at increasing heights above ground level, based on Professor Stokes' work *On the Effect of Wind on the Intensity of Sound*.

[7]   Jonathan Hulls, a farmer from Chipping Campden in the Cotswolds, realised the potential of the low pressure steam engine and in 1736 demonstrated his prototype paddle steamer on the River Avon at Evesham.

attached to the chimney. Boulton and Watt became involved in developing marine engines and the size of steam vessels and the number of steam-powered Atlantic crossings increased in the 1820s and 1830s. In 1838 the *Great Western* steamed into New York at over ten knots. The *Great Britain*, launched five years later, was screw-driven[8] by 1200-horse-power engines, but paddle wheels remained more common than screws for many years into the future.

Despite the advances in steam-powered vessels, sailing ships formed 90% of the total tonnage of the British merchant fleet in the 1860s. Steam vessels were less dependent on the weather but were exposed to the same danger in storms and they also suffered from the similar dangers of poor construction and overloading experienced by sailing vessels. Collier vessels plying the North Sea coast were particularly unseaworthy.

Lloyds figures from 1869[9] listed British shipping losses as 192 steam ships, 98 ships, 796 schooners, 468 brigs, 327 barques, 265 brigantines and 177 smacks and smaller craft.[10] Of this total 660 vessels were lost in moderate to strong breezes and 177 lost in calm weather. Thirteen vessels were lost at anchor when steamers ran them down in clear visibility! Eight vessels were lost through collision in fog but it is unclear from the records if fog was a contributory factor in any of the other losses.

Vessels at anchor were obliged to sound their bells every minute in poor visibility and Admiralty Rules of Navigation later standardized other sound signals. On sailing vessels it was the usual practice to lay up in fog as the least dangerous option;

---

[8]   A steam tug propelled by screw rather than paddles was trialled on the Thames in the 1830s, designed by John Ericsson and screw propulsion was patented by Ericsson in the USA in 1836. John Ericsson, incidentally, submitted his steam locomotive *Novelty* for assessment in the famous Rainhill Trials along with Stephenson's locomotive *Rocket* in 1829.

[9]   Close examination of Lloyds' figures by Samuel Plimsoll showed that nearly 56% of shipwrecks around Britain were colliers lost on the east coast. These losses cost the lives of 850 men each year between 1861 and 1871. There was a huge trade to ship coal from the north-east of England to London and from the 1820s many new harbours were developed in the north-east to export coal. Collier trade was lucrative for the collier owners (who owned the mines in many cases), who succeeded in blocking legislation on the overloading of old vessels, until a violent storm in 1871 led to a great loss of life off the coast near Bridlington.

Plimsoll submitted evidence to Parliament which showed that great stresses were exerted on a vessel when it was deep in the water as a result of overloading and it was a fact that poor construction and maintenance of older wooden vessels led to a situation where many boats simply fell apart under the strain. Ship owners were adequately insured against such losses and did not appreciate the need for improvements in safety at sea. The underwriters were only individually liable for small amounts despite helping to provide cover for many thousands of pounds and they were also sufficiently astute to reinsure against their own risks! The high costs, lost goods, insurance cover and seamen's lives did not encourage safer practice but legislation finally did.

[10]   Astonishingly, these huge losses represented only a very small percentage of the British merchant fleet, which by the turn of the century exceeded the total shipping of all other nations put together. Lloyds' Register of Losses continued to indicate an average of 1–1½% tonnage lost each year. Into the new century the Register showed that sailing ship losses were over three times greater than steam ships.

navigating without any visible points from which to obtain a position fix enforced a reliance on dead reckoning and frequent use of the lead.[11]

Captains of steam ships had a better idea of their position in fog if they could keep going at a constant speed to minimise the effect of wind and tide and gain relatively greater accuracy in plotting the position on the chart, based on the soundings and dead reckoning. This encouraged captains not to slow down much in fog, and in any case it was fatalistically acknowledged that it was better to risk running into another vessel than to be run into while stationary!

'To lie to until the fog clears off is the safest course to pursue but passengers go and shippers send their goods by those lines which have the reputation of making the quickest voyages, … hence owners … press their captains significantly to waste no time and any over-cautious mariner runs the risk of being told to seek another berth'.[12]

The shipping and finance journal, *The Siren*, as late as 1896 included an editorial which asked how trade could continue in Britain if a ship was obliged to drop anchor every time it became foggy.[13] As the percentage of steam vessels increased, shipping became less dependent on the weather and was encouraged to run against the competition in all conditions. Commercial pressures demanded larger and faster vessels and lobbying began for coastal fog signals to protect shipping in poor visibility when lighthouses were ineffective.

The recognition of the growing need for fog signals was accompanied by some concern that their implementation might actually increase the danger to shipping by encouraging mariners to proceed in fog.

The steam whistle was invented in the early 1830s as a warning device to indicate excess pressure building up in a boiler. No patent was taken out on the device and its obvious usefulness resulted in its widespread manufacture and application. Whistles became familiar at factories, on the railways and aboard shipping. At sea the steam whistle supplemented the bell and was used to indicate the movement of a vessel in fog.[14]

---

[11] A position fix requires accurate observation of stars or points on the coastline. To help establish a vessel's position without these observations, the course being steered and the speed of the vessel through the water is plotted on a chart to give a dead reckoning. This course is adjusted to allow for the effect of the current and leeway to obtain an estimated position. The 'lead' line indicates the depth of water below the vessel - knots in the rope show each fathom. If position is correct the depth of water measured by the lead should correspond to the submarine contours shown on the chart.

[12] Paper presented in Edinburgh in 1881 by D A Stevenson: 'On Coast Fog Signals'. D.A. Stevenson was Engineer to the Commissioners of Northern Lighthouses from 1885 to 1936.

[13] The editor of *The Siren* believed he was speaking on behalf of shareholders who had £125,000,000 of investment in British and colonial shipping.

[14] From the haphazard use of signals, by 1880 there were international agreements in place which required steam vessels to sound a whistle when proceeding in fog: One prolonged blast if on the starboard tack and two if on the port tack. By 1890 Article 12 of the Rules of the Road at Sea insisted on a prolonged blast to be sounded by vessels under way in fog at not more than two minute intervals. The 1897 *Regulations for Preventing Collisions at Sea* also insisted vessels should only '… go at a moderate speed … in fog'.

In 1845 Alexander Gordon, a member of the Institution of Civil Engineers, gave evidence to the Select Committee on Lighthouses suggesting the use of a whistle with a reflector as a coastal fog signal. He thought that the instrument could be sounded either by steam or by 'condensed air'. Admiral J.N. Taylor had submitted plans to the Admiralty in the previous year for a signalling device which was intended to assist ship-to-ship communication over a distance. Taylor's proposal appears to be the earliest record of a mechanical air-operated reed signal. His fog trumpet would produce sound in four pipes of different lengths and, not unlike a small church organ, the instrument would have had a hand-cranked pump to deliver the 'condensed air'. The success of his communication reed horn is not known but the name Taylor gave his instrument is familiar: his 'telephone' predates the development of Bell and Edison by over thirty years!

In 1860, Mr E.A. Cowper, a Fellow of the Royal Society, suggested to the Elder Brethren of Trinity House that a speaking trumpet such as was used for ship-to-ship communication,[15] could be fitted to a steam whistle. This fog trumpet would be on a large scale, perhaps up to twenty feet long, and Cowper thought that if sounded at a pressure of about seventy pounds it might be effective for up to ten miles. To provide the steam at relatively short notice, Cowper suggested that the boiler should be specifically designed with a large surface area for a small amount of water. His proposals were not acted upon by Trinity House. Cowper also suggested that a distinctive character should be sounded by a fog signal.

In Nova Scotia on 22nd October 1854 an article in *The Morning News* of New Brunswick announced the invention of the steam-whistle fog signal by Robert Foulis. Foulis had apparently heard his daughter practising the piano as he walked home on a foggy night and noted that the higher frequencies were faint although the low notes penetrated the fog to much greater distances. These observations led to the development of a low note whistle which was finally placed in the harbour at St John's in 1857. The large whistle used super-heated steam to produce its note, whereas whistles used aboard steamers were limited in their effectiveness by the low pressure steam boilers.

Foulis' instrument was developed by Vernon Smith who sold on the patent in this country to a Mr Barker in 1861. This development of the steam whistle made it capable of sounding a blast automatically by means of a clockwork-actuated valve. Another variation was primarily intended for use at sea which enabled the whistle to be manually controlled to send messages using Morse or other codes.[16]

The steam-whistle 'fog alarm' was introduced at a number of sites in North America but few were used on the British coast. The Commissioners of Northern

---

[15] Samuel Moreland is credited with introducing the speaking trumpet for ship to ship communication in the 1680s.

[16] In 1876 Captain Richards of the Admiralty Hydrographic Office proposed a fog signal code for ship-to-ship and ship-to-shore communications using a ship's whistle.

Lighthouses undertook whistle trials in Scotland and a six-inch whistle was subsequently installed at Girdle Ness lighthouse near Aberdeen in 1869.[17]

Two whistles were introduced at Cloch Lighthouse on the River Clyde and remained in use for nearly thirty years:-

> '...*two steam whistles, the first low, the second high, give four blasts at equal intervals every half minute*'.[18]

Seamen who regularly crossed the North Atlantic often had first-hand experience of the American signals and reported favourably on the usefulness of the whistles in fog. This led some ship owners to lobby the lighthouse authorities and the Board of Trade, but caution and deliberation by the lighthouse authorities recognised that alternative types of signal were also being developed.

Lighthouse engineers and consultants to the lighthouse authorities attended the demonstration of Daboll's reed fog signal at Dungeness in 1862. After the demonstrations David and Thomas Stevenson, (who recommended the two-note steam whistles installed at Cloch) recommended to the Cumbrae Lighthouse Trust that a Daboll reed should be installed at Little Cumbrae Lighthouse and the first reed fog signal in Scotland was installed at Cumbrae in 1865.

Celadon Daboll conducted many experiments for the US Lighthouse Board in the 1850s and developed mechanical fog bell machinery for use in several lighthouses. In 1851 he established trials at Beaver Tail Point Lighthouse, Rhode Island, using a trumpet attached to a locomotive whistle. Later that year, also at Beaver Tail Point, he developed his reed horn which was sounded by compressed air. A 17-foot high trumpet was attached to a reed box mounted directly on to an air receiver. Two pumps were worked by a horse to compress the air to the sounding pressure of 40 lbs per square inch (psi). Daboll also tried reduced pressures and different types of reed. The original steel reed was about 10 inches long and 2 inches wide and tapered to an eighth of an inch at the free end. Following the trials, the standard length was later reduced to 5 inches.

When the air under pressure in the storage receiver was admitted to the reed box, initially by using a hand-operated valve, the free end of the reed was caused to rapidly vibrate as the air escaped. The high speed opening and closing of the oscillating reed produced separate bursts of sound in the trumpet. The instrument was improved over the following years and a more practical design later replaced the horse with a small hot air engine to compress the air. A system of connecting rods and wires was also developed which enabled the length of the blast and the period of silence to be automatically controlled by a drive from the engine. This

---

[17] Thomas Stevenson carried out experiments with a steam whistle in 1865 which incorporated an iron holophone reflector. Directional improvement increased the range of the signal but not by a significant factor.

[18] Cowper, F. *Sailing Tours*, 1895.

regulating system also enabled the signal to be operated at a distance from the engine and compressor if necessary. A further development was to mechanically rotate the trumpet so that the sound would be distributed over a wide area.

This instrument consolidated many of the early developments in fog signalling machinery and Daboll took out a patent on the use of compressed air horns in America (1860) and in Britain (1863).

The equipment demonstrated at Dungeness in 1862 had a reed which was sounded at about 10 psi and the trumpet was 4 feet 6 inches long with a 21-inch mouth and a 2½-inch diameter throat. After the demonstrations, this equipment remained in use at Dungeness until 1865 when it was replaced by a new Daboll trumpet with a 9-foot long horn. The trumpet had originally acted as a conductor of the sound but a further improvement to the power of the instrument was produced by tuning the reed to the frequency at which the horn vibrated.

The first trials with the original reed used a stationary trumpet positioned horizontally 'eight feet above the shingle on the beach.'[19] A four-second blast was sounded with twelve-second intervals but the results were variable. Off axis, out at sea, the noise of a steamer's paddles was sufficient to render the horn inaudible to observers, even quite close by. Peculiarly, on other occasions, the signal could be heard distinctly. Eighteen-pound cannon were subsequently introduced for comparison with the range of the reed. Perhaps a more powerful reed instrument would improve the range.

The new first order Daboll horn was installed in 1865 at Dungeness at a cost of £1,238 and the third order original instrument was subsequently removed to the Newarp light vessel in 1868. The new instrument had the further improvements of a curved, vertical trumpet and the automatic rotating mechanism. The reed was sounded at a higher pressure, 12 psi, falling to 5 psi during the blast.

A larger Ericsson hot air engine was obtained, although the machine was still only 2½ horse power. Trinity House had two engines installed, a service engine and a standby. The new signal at Dungeness was included in the 1866 edition of the Admiralty List of Lights:

> *'Daboll's fog horn is sounded during foggy weather. The mouth of the horn will traverse an arc of two hundred and ten degrees, viz; from NE by E, E, round to South to W. N, and vice versa, so as to point in every direction between those bearings once in each minute; the duration of the sound being five seconds, with an interval of twenty seconds between each blast.'*

Dr Tyndall described the same equipment in 1873 in his Report on Fog Signals. The trumpet still rotated automatically through 210 degrees:

> *'halting at four different points on the arc and emitting a blast of six seconds duration ...'*

---

[19]   Phonic Coastal Signals, *Minutes of the Proceedings of the Institution of Civil Engineers*, March 1871.

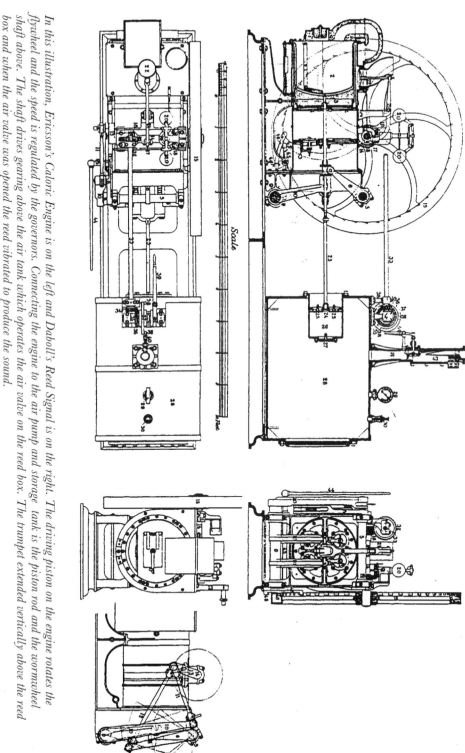

In this illustration, Ericsson's Caloric Engine is on the left and Daboll's Reed Signal is on the right. The driving piston on the engine rotates the flywheel and the speed is regulated by the governors. Connecting the engine to the air pump and storage tank is the piston rod and the wormwheel shaft above. The shaft drives gearing above the air tank which operates the air valve on the reed box. The trumpet extended vertically above the reed box and when the air valve was opened the reed vibrated to produce the sound.

He noted that the sounding pressure started at 8½ psi and fell to 5¾ psi.

Instructions to Lightkeepers, issued to the US Lighthouse Service, prescribed a sounding pressure of between 8 and 11 psi for the Daboll reed.

The same Ericsson caloric engine was still in use in 1873 when the Engineer in Chief to Trinity House, James Douglass, confirmed that the engines had been totally reliable, despite the poor reputation of hot air engines in general.

> The early hot air engines were weak and primitive but the many improvements had resulted in a practical machine. The hot air engine used extremes of heat to drive a piston and this energy was converted into a rotary motion by a flywheel. One of the earliest practical engines was the Stirling engine of 1827, but the invention is credited to Sir George Cayley who later patented a design in 1837 which developed from his 1807 engine. The basic principles of Cayley's engine were incorporated in the Buckett engines later used at some fog stations in the 1870s. Coal or coke was used to heat the air and air under pressure was pumped into the furnace before admission to a vertical cylinder where it expanded and forced a piston to rise, which also operated valves to control the flow of air. Exhaust hot air from the cylinder was directed through the 'regenerator' to heat the cool air intake and the intake was additionally warmed by compression en route to the 'regenerator' and thence into the furnace. The hot air in the furnace was then directed into the main cylinder at the bottom of the piston stroke.
>
> There were several different types of machine available which operated by using variations of the above cycle, but the most efficient engines used a 'regenerator' of some kind and had an internal heat source capable of operating at a high pressure. The Wilcox engine, from 1862, used a rotary valve linking two adjacent cylinders, to drive a crank shaft at 120 revs per minute. The operating temperature of the hot air was 600 degrees Fahrenheit and *The Engineer* was confident that the machine was quiet, smooth running and 'powerful and economical in fuel and attendance.'

The Swedish inventor John Ericsson first demonstrated his hot air or 'caloric' engine in England in 1833, but he continued to develop the engine in America into the 1860s. One of his engines was demonstrated at the Great Exhibition of 1851. His single cylinder engine comprised an inner cylinder and outer piston; cold air was drawn into the cylinder and compressed before being fed into an inner piston where the temperature and pressure were further increased. A weighted flywheel carried the engine over to the next stroke. Ericsson's engines were quite widely used in America and were manufactured after 1874 by A. & F. Brown of New York.[20]

---

[20] Ericsson's patent expired in 1870 and his design became widely used. A. & F. Brown produced Ericsson's engines in various sizes from ½ hp to 6 hp. The smallest engine had a 9-inch cylinder and cost $250. The 6 hp engine had a 32-inch cylinder and drove a flywheel at 35 rpm. This engine cost $2,250 in 1875. Ericsson engines, like the Stirling and Buckett engines, heated air in contact with the iron retort's sides. The engine developed by Browns passed the air under pressure through the furnace itself.

In the UK, Professor Holmes and the Caloric Engine & Siren Fog Signal Company held the patent for the Buckett engine and in the 1870s these engines were used for compressing air for the fog signals developed and manufactured by Holmes. The Buckett engines divided the air drawn into the machine along separate routes, the first passed through the fuel, gas coke, where the oxygen combined with the carbon to form carbonic acid and then carbonic oxide. The other route drew air over the fuel combining the gas mixture to produce an intense flame which, reputedly, was so efficient in consuming all of the carbon that no chimney was required! Nearly 12 horse-power was developed by these engines, which burned two pounds of coke per hour.

Professor Holmes' involvement with fog signal equipment is first noted in the early 1850s. He had close relations with Trinity House over many years, specifically relating to the introduction of electric illumination for lighthouses.

Holmes lived and worked in France between 1853 and 1856 developing the electric magneto with Monsieur F. Nollet, who patented it. Holmes may well have had connections with M. Reynaud, who presided over the public works ministry with responsibility for lighthouses. In any case, Holmes' 1863 reed trumpet was trialed in France in comparative tests with a whistle and bell in 1865. Holmes' reed was five and a half inches long, one inch wide and five sixteenths of an inch thick and made of 'German silver'. The signal sounded at 30 psi and the trumpet was designed to be tuned with the reed and carefully constructed to give even sound distribution throughout a total angle of 90 degrees. Holmes also produced a lower pressure reed horn which had a wider reed of 2½ inches by 6½ inches in length.

James Douglass, the Trinity House Engineer in Chief, was impressed by the Holmes' reed and encouraged the development of the instrument for use on light vessels. A similar instrument was subsequently installed on the Seven Stones Lightvessel off Land's End in 1871.

Another Holmes' reed was established about the same time at the new lighthouse at Souter Point. This signal had two trumpets set at right angles to cover 180 degrees of arc. The signal operated automatically when the pressure of the compressed air reached 30 psi and a tumbler was thrown over to allow air through to the reeds. As the pressure dropped during the blast to 28 psi, the weighted tumbler toppled back to cut off the air supply.

Rather mysteriously, the Admiralty List of 1872 says that the trumpet at Souter: 'rotates through 180 degrees of arc', but by this time, James Douglass was already of the opinion that the rotating horn, designed to give a wide spread of sound, was ineffective and led to unpredictable attenuation of the sound. He thought, with Holmes, that the use of fixed trumpets to cover a specific arc was a better arrangement. Nevertheless, at Dungeness the Daboll trumpets continued to rotate until 1875.

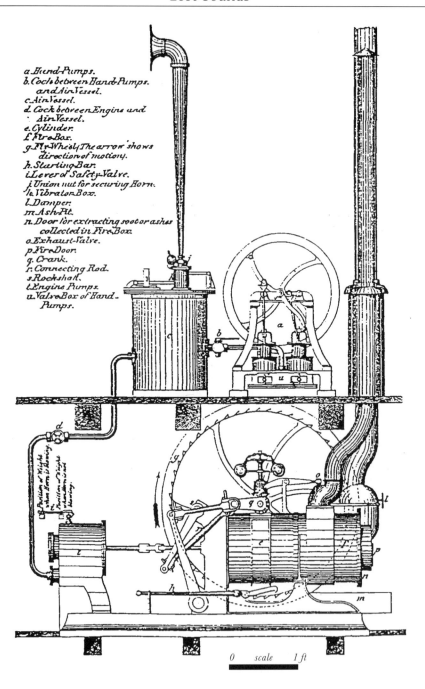

a. Hand-Pumps.
b. Cocks between Hand-Pumps
    and Air Vessel.
c. Air Vessel.
d. Cock between Engine and
    Air Vessel.
e. Cylinder.
f. Fire Bar.
g. Fly Wheel. The arrow shows
    direction of motion.
h. Starting Bar.
i. Lever of Safety Valve.
j. Union nut for securing Horn.
k. Vibrator Box.
l. Damper.
m. Ash Pit.
n. Door for extracting soot or ashes
    collected in Fire Box.
o. Exhaust Valve.
p. Fire Door.
q. Crank.
r. Connecting Rod.
s. Rockshaft.
t. Engine Pumps.
u. Valve Box of Hand-
    Pumps.

0   scale   1 ft

Holmes' fog horn as installed on the Seven Stones Lightvessel in 1871 (see Ch. 10). Below deck is the caloric engine and pump which supplied air to the receiver on deck. Above the receiver is the trumpet and reed box and the horn sounded when air pressure tripped the weight-operated valve. Next to the air vessel on deck are the hand pumps used when the engine was not operable.

Only a few Daboll reeds were ultimately in service in Britain[21] and despite the innovative design, the reputation of the instrument was not great in North America, where it was more widely used:

> *'... the trumpet was expensive to maintain, unreliable in working and liable to break down when most needed'.*[22]

Professor Tyndall was more generous in describing the Daboll reed at Dungeness:

> *'It is a fine instrument and its application was ably worked out by its inventor; ...(but) ... it would require very favourable atmospheric conditions to enable it to warn a steamer before she had come dangerously close to the shore'.*[23]

The reed signal fared better in Britain because of the developments of Professor Holmes. Several other manufacturers later made reed fog signals, including Chance Brothers. Reed signals were relatively cheap to produce and operate as they were sounded by low pressure compressed air. They produced a note of between 400 and 500 hertz and were usually audible for several miles.

In addition to their use aboard lightvessels, reeds were installed at harbour lights[24] and some major lighthouses into the early 1900s, where they remained, with periodic refurbishment, for many years. Replacement reeds were routinely manufactured by apprentices at the Trinity House Engineering Depot at Blackwall.

Wolf Rock, Skokholm and Needles Lighthouses all had reed fog signals installed in the early 1900s which survived until after the Second World War. Reeds at other lighthouses, such as Hartland Point and South Stack were replaced by more powerful compressed air signals in the 1920s and 1930s. The twin reeds at Penlee Fog Signal Station near Plymouth, installed in 1902, remained in service until the 1970s. As recently as the early 1990s the Admiralty List of Lights referred to other reeds still in use at a few harbour lighthouses.[25]

The Daboll reed at Dungeness, like the other early instruments, was not long lived and it was replaced in 1875 by a new fog signal, a gift from the United States. A new building was specifically constructed, from the end wall of which protruded a long horizontal trumpet which rested on a timber framework. This fog signal is believed to have been the original siren used in the fog signal trials conducted at South Foreland lighthouse in 1873 and 1874.

---

[21]   Daboll reeds used in British waters included Dungeness, 1862 and 1865; St Catherine's, 1868, Newarp 1868 in England; Cumbrae in Scotland, 1865 and Howth Bailey in Ireland, 1871.

[22]   Quotation from Canadian Lighthouse Authorities in Talbot, Frederick *Lightships and Lighthouses*, 1913.

[23]   Tyndall, John, 'Report to the Trinity House upon recent experiments with regard to Fog Signals', 1874.

[24]   At Dover '... there is a fog horn of the Reed type, recommended by the Trinity House authorities, and supplied with air by a compressor worked by an electric motor in a building alongside the lighthouse tower. The working air pressure for the horn is 10 lb per square inch.' *Engineering*, 8th July 1904.

[25]   Though still listed as 'reed' signals, the last survivors were replaced by electric emitters of similar frequency in the late 1980s and early 1990s.

Reed equipment similar to that installed at the Needles Lighthouse in 1906. Chance Archive, Courtesy Pilkingtons.

The original 1875 fog signal building and its small 'low light' was modified and two-note sirens with bell-mouthed trumpets were added in 1904. The corrugated iron building was subsequently replaced by a brick one with an iron tower for the 'low light'. Courtesy Trinity House National Lighthouse Centre.

Chapter 3

South Foreland

and the fog signal trials

By the mid 1860s there were 2814 lighthouses protecting the coasts of the world with 1785 on the European coasts.[1]

Around Britain there were 357 lighthouses[2] of which 160 were maintained by local port and harbour authorities and the remainder were operated by the three lighthouse authorities, the Commissioners of Northern Lighthouses, the Commissioners of Irish Lighthouses and Trinity House, which was responsible for the lights of England and Wales.

Lights on the British coast were generally lit by oil lamps at this time[3] but there were experiments to evaluate the brightest, cleanest and most cost-effective illuminant. With the development of the gas industry from the 1830s some harbour lighthouses were linked to the gas main and lit by town gas, and by the 1850s experiments were under way into using electric arc lighting. South Foreland became the first electric lighthouse in 1858, less than 40 years after the last of the open coal braziers had been discontinued at St Bee's Head.

Most of the lights shown were 'fixed', that is, were continuously shown, and to help the mariner distinguish one lighthouse from another, some showed two or more lights. Twin towers remain at St Ann's Head, Portland, the Lizard, Nash Point, South Foreland and elsewhere.[4] These lights burned oil but had all been lit by coal braziers.

---

[1]   Europe: 1,785. America: 674. Asia: 162. Oceana: 100. Africa: 93. *Engineering*, 7th February 1868.

[2]   This figure increased to nearly 500 within a decade.

[3]   Colza was a vegetable oil which replaced sperm whale oil.

[4]   Twin towers had another function at some sites, which was to act as leading lights. If leading lights were kept in line by an approaching vessel this indicated the optimum course for a vessel to steer. The light could warn of shoals or reefs. South Foreland lights in line indicated the southern extremity of the Goodwin Sands.

Other distinguishing features included the use of coloured lights but the most modern lighthouses screened off the light for a regular fixed interval.

The Fresnel lens was introduced in the 1830s but it was towards the end of the 19th century that the powerful dioptric and catadioptric group flashing optics were developed which enabled distinctive characteristics to be visible at great distances.[5] Prior to the development of these wonderful lens systems, however, lights were categorized in the Admiralty List of Lights as fixed, coloured or intermittent. The early revolving mechanisms could take several minutes to rotate and to the mariner the effect was a gradual dimming and increasing of the brightness. If the chart listed a 'flashing light' this indicated a frequency in excess of five flashes per minute. An occulting light had shorter periods of darkness than light.

Of the different types, fixed and revolving lights predominated at the time. The problem with fixed lights was that they could sometimes be difficult to distinguish from other lights on shore or on other ships, and the problem with the revolving lights was that it was difficult to guarantee a regular period and the slow rotation could be almost impossible to detect from a moving vessel in poor weather.

The mathematician Charles Babbage gave some thought to maritime safety and suggested that all lighthouses should show fixed lights and have occulting mechanisms installed which would operate by clockwork to eclipse the light automatically to produce a very specific character. The identity of each lighthouse could be specified by a number on Admiralty charts and the occulting light would repeat its own number between intervals: lighthouse 243, for example, could have 2 occultations followed a pause, 4 occultations then a pause, then 3 occultations and a longer interval. Babbage thought that the same principle could be applied to the sounding of fog signals, although in the mid-1850s, when he suggested this idea, it could only have applied to bells.

Babbage thought a sound signal could identify a light vessel by using a similar numerical system to that proposed for the lighthouses so that a seaman might know the nature of the danger he was warned of by the sound. He realized, however, that the effective range of the bell was very poor in comparison with the lights:

> *'It is unfortunate that the means of warning the seaman of his danger should extend to the shortest distance when the danger is most imminent'.*[6]

and he was disappointed that so little research had been undertaken into sound producing signals and their audible ranges.

---

[5]   From the mid-1830s revolving and static refracting lenses were installed at new lighthouses and gradually replaced the Argand lamps and reflectors at established lights. The dioptric lenses gave a brighter output from only a single lamp, unlike the catoptric apparatus which required oil lamps for each reflector. Dioptric lenses used rings of prisms to *refract* light into concentrated beams. Catadioptric lenses used supplemental prisms above and below the refracting elements to *reflect* stray light and maximise output.

[6]   Babbage, Charles, 'Notes respecting lighthouses', 1852.

Babbage was aware of experimental work with fog trumpets and thought that there should be research into the nature of the most suitable and effective sounds. Perhaps low notes would be lost in the noise of breaking waves and perhaps a whistle would be smothered by the sound of the wind blowing through the ship's rigging. He wondered if multiple discordant notes might be the most effective sound for a fog signal. The most disagreeable sounds that could be produced would certainly be the most noticeable![7]

Babbage wrote to the US Lighthouse Board besides the British authorities and although his suggestions did not lead directly to any specific developments they added to the growing opinion in favour of thorough experimentation with signalling devices.

Alexander Beazeley presented a paper to the Institution of Civil Engineers which berated the scant research into sound transmission in fog and the development of fog signal technology, and he urged for thorough experiments to be undertaken. He also endorsed the idea of correspondence between the character of the light and the character of the fog signal. For instance, a fixed light might correspond to a continuous sound blast; a flashing light might have corresponding intervals in the sound signal; and where a white and coloured light was shown the equivalent sound signal might have two separate notes. Some people wished to make the connection between the light and the sound even more specific with an equivalence between the revolving light and the sound emanating from a revolving trumpet.[8]

Sir William Thompson, who later became Lord Kelvin, thought that Babbage's numerical system could be improved upon by the use of Morse code with an eclipsing light so that each lighthouse would proclaim its identity to the mariner.[9] Opponents of the idea believed that accurate observation and decoding of the signal would be impractically difficult, even for the most capable mariner.

Michael Faraday was the Scientific Advisor to Trinity House until 1866 and sought extensive experimentation with fog signals and an enquiry into the logistics of their implementation. He was strongly opposed to an uncoordinated, piecemeal approach to installing signals, and believed that the unproven and possibly exaggerated claims made for the ranges of certain signals, generally North American whistles, could be dangerously misleading. He wanted to be sure that a signal would

---

[7] Babbage also thought a parabolic reflector would help to concentrate sound and wondered if underwater signals should be tested.

[8] This was not the reasoning behind Daboll's rotating trumpet which was designed to give the widest spread of sound.

[9] By the early 1880s, James Douglass reintroduced the idea of using Morse code with fog signals. By using high and low blasts, Morse code could spell the initial letter of each fog station. The idea was not implemented but Morse characteristics were used and remain in use at some stations. There is no significance attached to the letters used, merely their distinctive character: e.g. 'N' (dash, dot) was used at the Lizard . Kelvin's idea was later used with radio beacons which repeated their call signs to aid position fixing from the 1920s and the Racon system uses a Morse identification on a ship's radar.

be effective over a given distance, and reliably operated, if it was to be installed and listed in Admiralty charts. He commented 'A false promise to the mariner is worse than no promise at all'.

His opponents did not agree that a false sense of security could be more dangerous and wanted sound signals all round the coast so that a ship could navigate by them, picking up each successive one as the ship moved up the coast, as it did with lighthouses on clear nights.

Alexander Cunningham offered a rationale for best utilizing the distinctive qualities of different types of fog signal at specific sites: he thought guns could be used at major landfalls, trumpets at bays and estuaries and bells at harbours.

In March 1873 a committee reporting to the House of Commons circulated a report on fog signals currently in operation, along with a supplemental document which contained details of a visit made by members of the Elder Brethren of Trinity House to North America in 1872. The report listed the signals in use in England and Wales at the time, which consisted of bells at 19 lighthouses, the Daboll reeds at Dungeness, St Catherine's and Newarp lightvessel and the Holmes reeds at Souter Point Lighthouse and the Seven Stones and Owers light vessels. Fog guns were then in use at North Stack, Flamborough Head and Lundy Island. Of the 39 lightvessels then in service on the English coast, the report stated, '...thirty-six are furnished with a gong'.

There was a similar picture in North America. Bells were the most common signal in use, but there were also nearly 30 steam and compressed air fog signals. The 'steam fog alarms' were 6, 8 or 10-inch steam whistles which were sounded at about 70 psi, and some had directional trumpets. It is possible that steam was used with some of the reed horns, but reeds were usually operated by compressed air. There were, in addition, two other steam-powered signals in use in the United States in 1871, one at Sandy Hook and the other at Staten Island. These were the first siren fog signals.

The House of Commons' report recognized that the Canadian steam whistle was cheap, simple to use and generally very effective and the 10 and 12 inch whistles had impressed the authors. The authors were less impressed by the Daboll reeds seen in America, and were particularly concerned about the lack of any back-up in case of mechanical failure: at one lighthouse visited the fog signal machinery lay idle after a breakdown. The report, therefore, demanded reliable machinery and recommended that a duplicate engine and compressor should be installed at each lighthouse, complete with spare reeds.

Fog prevailed on the English coast, on average, for less than 70 days each year,[10] whereas the east coast of America had much more fog and the need for fog signals

---

[10]   The figures come from unpublished Trinity House records from North Stack, Lundy and Flamborough Head, where the number of foggy days in the year to June 1872 were 53, 67 and 68 respectively.

One of the Brown sirens in use near New York about 1872. The vertical siren disc inside the trumpet is mounted above the steam boiler. Adjacent to the fire-box and boiler is the small steam engine which spins the siren via a connecting pulley from the drive shaft. The engine also controls the valve which admits steam to the siren. Courtesy The Engineer.

was more obvious. The report, nevertheless, recommended further development of the use of fog signals in Britain. The Daboll and Holmes reeds, already in service, were particularly recommended for use at harbours although the widespread installation of fog signals along the coast was not thought to be a good idea. The erection of new signals at major lighthouses was recommended including Portland and the Lizard and at several lightvessels, and reeds were to be subsequently considered for installation at Hartland Point, St Ann's Head, Longstone, Spurn Point and Flamborough Head.

The authors of the report, Sir Frederick Arrow, the Deputy Master of the Trinity House, and Captain Sydney Webb RN, had been accompanied by other individuals on their inspection of the United States' lighthouses in 1872, including Dr John Tyndall. Tyndall joined them after his American lecture tour on the subject of sound.

Professor Joseph Henry, the chairman of the US Lighthouse Board, and General Woodruff, the district engineer in charge, accompanied Tyndall on his visit to the lighthouses at Staten Island and Sandy Hook to see the steam 'syren'. Tyndall was impressed by the sirens and subsequently visited the Progress Works in New York where Adolphus Brown demonstrated the equipment more fully.

> *'In the steam syren patented by Mr Brown ... a fixed disc and a rotating disc are ... employed, radial slits being cut in both the discs ... One disc is fixed across the throat of a trumpet 16½ feet long, 5 inches in diameter, where the disc crosses it, and gradually opening out 'till at the other extremity it reaches a diameter of 2 feet 3 inches. Behind the fixed disc is the rotating one, which is driven by separate mechanism ... When the radial slits of the discs coincide, a puff of steam escapes. Sound waves of great intensity are thus sent through the air; the pitch of the note produced depending on the rapidity with which the puffs succeed each other; in other words, upon the velocity of rotation'.*[11]

Steam at 70 psi directed through the siren was interrupted by the movement of the disc to produce a series of short bursts from the continuous stream admitted. The rapid succession of these bursts set up vibrations in the trumpet to produce a note of about 560 Hz.

It is possible that Professor Tyndall may have illustrated his American lectures on sound with a device which he first demonstrated in his Royal Institution lectures in the early 1860s.

The invention of the 'syren' in the early 19th century is credited to Dr John Robison, though developed by Cagniard de la Tour. The instrument Tyndall used in his laboratory was made of brass and steel and was used principally as a device to measure the frequency of various sounds.[12]

The sound produced by the 'syren' resulted from the puffs of air which escaped from corresponding holes in a stator and spinning rotor. The laboratory siren had sixteen adjustable holes in the rotor and stator and by varying the ratio of holes in both parts of the instrument,. different properties of sound could be demonstrated, such as, for example, a fundamental note and its octave.

Tyndall's laboratory siren was a self-rotating instrument as developed by the German physicist Helmholtz. The holes in the siren were round and cut at opposing

---

[11] Tyndall, John, 'Report to the Trinity House upon Recent Experiments with regard to Fog Signals', 1874. Except where indicated subsequent quotations are also from the report.

[12] Tyndall demonstrated how the siren could measure the frequency of a tuning fork: when the note of the siren corresponded to the note of the tuning fork a counter on the siren was activated and timed for one minute. If the counter indicated the siren had completed say 1,440 revolutions, this figure was divided by 60 to give the number of revolutions per second - 24. The siren had 16 apertures in the rotor and stator so multiplying 16 by the number of revolutions per second, 24, resulted in the total 384 (Hz) the frequency of the note – about F sharp. The wavelength of the tuning fork, having a frequency of 384 vibrations per second, could be calculated by dividing the speed of sound in air by the frequency of the note (1120 divided by 384) to arrive at a wavelength of 2 feet 11 inches.

angles in the rotor and stator, so that air impinging on the opposing surfaces forced the rotor to spin: the escaping air drove the siren round and produced the sound.

In the steam siren made by Browns, the rotor was motor-driven and the apertures were cut radially in the rotor and stator discs.

Tyndall was very impressed by the practical application of the siren as a fog signal. American trials of the siren had proved very favourable but Tyndall was keen to conduct his own experiments to build on his theoretical work and consolidate the performance of the siren relative to other fog signals. He proposed a series of fog signal trials which would compare the effectiveness of different signals in different meteorological conditions over a period of months, using observers on land and at sea.

The Elder Brethren of Trinity House encouraged their Scientific Advisor to additionally include guns and whistles in the fog signal trials but Tyndall was somewhat reluctant; he did not believe the claims that whistles were frequently heard at 20 and 30 miles distance.

Dungeness, unusually, did not host the fog signal trials and instead South Foreland was chosen because of the high cliff afforded. Signals were to be installed at the top and bottom of the cliff to compare the effectiveness of sound transmission from the two positions.[13]

The preparations made at South Foreland incorporated assembling and installing the necessary equipment, including upright steam boilers, air receivers and an Ericsson caloric engine such as that used with the reed fog signal at Dungeness. The steam engine at South Foreland used to generate electricity for the arc light was also used in some of the trials.[14]

On the cliff top at the start of the trials in May 1873 there was a timber construction supporting two brass trumpets,[15] a 6-inch whistle and a 12-inch whistle. (The siren was not included in the first trials but arrived some weeks later from Messrs. Brown courtesy of the US Lighthouse Board.) With the exception of the 12-inch whistle all of the other original instruments were sounded by compressed air at 18 psi. The 12-inch diameter whistle was 18 inches high and was sounded by steam at 64 psi.

---

[13] Following Professor Stokes' theories on the effects of wind on sound transmission, Tyndall believed that the increasing speed of the wind at increasing heights above the ground might produce the effect of forcing sound upwards in a facing wind. If this was the case, the range of a fog signal could be increased by installing it as high as possible and directing the sound downwards into the wind.

[14] The steam engine was also used to drive compressors to supply air to the fog signal. Although a steam engine at Souter Point performed this function, it was more efficient to use hot air engines to provide compressed air or, perhaps, use steam for the fog signal directly rather than use the steam engine for compressing air. With very few exceptions first class fog signals in Britain were operated by compressed air produced by hot air engines and later by oil and diesel engines.

[15] The reed fog signals were all Holmes' instruments with vertical trumpets 11 feet long and 22½ inches in diameter at the mouth. The reeds were 9 inches long, 2 inches wide and ¼ inch thick.

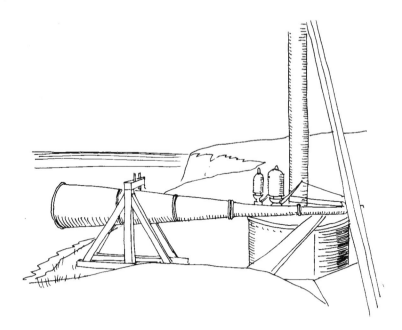

The Brown 'syren' trumpet and other fog signals assembled on the cliff top for the South Foreland trials.

Excluding the steam whistle, similar compressed air equipment was installed at the foot of the cliffs, all of which had to be carried down twelve ladders in the shaft that led to the bottom. The specially excavated lower site was 40 feet above high water and the cliff top site was 235 feet above high water.

The purpose of the trials was to establish if any signal could be consistently relied upon to be audible at four miles in all conditions, but other issues to be resolved were concerned with the type of sound and the means of its production: what were the relative merits of different types of signal? Which specific notes most favourably penetrated different types of fog? What frequency of sound would prove most consistently penetrative in all conditions? Should a fog signal be placed near the water or at the maximum possible height to enhance the range? Should a horn be turned to face into the prevailing wind? How did the sounding pressure affect the range of the signal? What were the maximum and minimum useful operating pressures for both air and steam?

If it was to be possible to navigate by the use of sound produced by fog signals, it was necessary to be able to gauge the direction of the sound source[16] and it was

---

[16] A dangerous but common practice pioneered by American sea captains was to try and ascertain the direction of the source of a fog signal and steam directly towards it, using the lead to estimate a position.

hoped the results of the trials would also shed light on this. However, it was already acknowledged that the apparent loudness of a signal was not a reliable indication of proximity to the signal and the attenuation of sound from a horn off axis was well known to many with experience of Daboll's rotating trumpet.

On May 19th the two Trinity House steam ships *Irene* and *Vestal*[17] with observers aboard, set off on predetermined routes at specific distances off the South Foreland headland. With some excitement the observers listened for each signal, but less than three miles out to sea 'with paddles stopped' the sounds of all signals were lost. At the end of the day the disappointed observers noted that the most impressive sound heard had been a cannon fired at Dover Castle!

Throughout the trials, the observations made on each day were collated and summarized. On the first day the sounds were all lost at three miles with the wind ENE 6 to 7, a barometric pressure of 29.9 inches and a temperature of 52 degrees Fahrenheit. The partial summaries of the daily tests, as recorded in Tyndall's report make interesting reading.[18] They are often contradictory and sometimes quite poetic, but they generally begin to take shape as Tyndall draws conclusions from the accumulating evidence.

Before the trials it was generally believed that conditions of poor visibility also produced poor conditions for sound transmission and Tyndall previously thought that light and sound energy would be equally absorbed:

> '... a portion of the vibration is reflected and lost ... at each of the innumerable surfaces where the air and globules of water ... touch ...'

Tyndall found that the whistles[19] expended great energy near the source but that sound energy rapidly fell away and he questioned the long audible ranges claimed for such signals in Canada.

Nevertheless, there were other recorded instances of sounds being audible over great distances and some of these had been made by reputable observers: Sir William Herschel, for instance, claimed to have heard cannons in battle over 200 miles

---

[17]   *Irene* was built in 1851 at Chatham Docks and registered as 149 tons. The paddle schooner was a wooden vessel with two thirteen-ton boilers and twin engines and was the first Trinity House ship to have steam paddles. *Irene* had a crew of 30. The *Vestal* was commissioned in 1855 and was used primarily for laying buoys and for Elder Brethren's annual inspection of lighthouses. This paddle schooner was registered as 199 tons. See Woodman and Dalton, *Keepers of the Sea*, 1983.

[18]   Each of the daily summaries concludes with a description of prevailing weather, wind speed and direction. The report describes any modifications to the equipment used from day to day and which instrument was being compared with which other. Resonators were swapped, sounding pressures varied, reflectors added or removed. Minimum and maximum ranges for each signal were recorded and for example the maximum ranges for all signals on the following dates were: 19th May, 3½ miles; 20th May, 5½ miles; 2nd June, 6 miles; 3rd June, 9 miles; 10th June, 9 miles; 25th June, 6 miles; 26th June, 9½ miles; 1st July, 12¾ miles; 2nd July, 4 miles; Tyndall 1874 *op. cit.*

[19]   Five different sizes of whistle were tested, made by British and American manufacturers. Despite elaborate screening and the use of reflectors – one 15 feet by 12 feet – the sound radiated in all directions, painfully loud near the source, but without great effect out at sea.

away. Far less dramatically, Alexander Beazeley had heard the Lundy fog gun at Hartland Point and Sir James Douglass heard Milford Haven guns 25 miles away at the Smalls lighthouse.[20]

The daily observations began to suggest that, contrary to expectations, there was poorer acoustic penetration on clear days than on days of mist and rain. During several days of heavy rain in October, all of the signals had improved ranges. Tyndall thought this was due to '... non-homogenous locks of air and vapour' having been 'abolished by the storm'.

If air pockets of variable density caused obstacles for sound transmission, then an homogenous atmosphere should be a better medium, and indeed one of the conclusions of the report was that foggy conditions presented the ideal medium for sound transmission.[21]

Fog does not only occur in calm weather but can sometimes be accompanied by strong winds. The deleterious effect of strong winds on the audible range of a fog signal was immediately obvious to the observers in the trials, but by the end of June they had noted that something other than wind affected the range, surprisingly so on clear days:

> '...` *days of the highest optical transparency proved themselves acoustically most opaque*'.

Aboard the *Irene* on a cloudless, hot day, Tyndall could hear no sounds from any of the signals, including the guns which had been introduced to the tests. Puffs of smoke from the inaudible guns were apparent as the *Irene* steamed to various positions. Sensing the sun beating on his back he realized that the local effect of the sun on different areas of the sea would produce differing rates of evaporation from the surface, causing rising air streams of variable density. These flocculant variations resulted in feeble and unpredictable sounds from the Foreland at best and silence at worst. In the evening, as the heat of the sun diminished, the sounds reappeared and, at 7 p.m., the intensity of the sound was 36 times greater than at 2 p.m., when two miles off the Foreland:

---

[20]   Research unavailable to Tyndall, since it was amassed during and after the Great War, determined that huge explosions at munitions factories and dumps were audible for about 60 miles. Transmission of the sound was not simple, however, and sound waves which radiated upwards through the atmosphere appeared to be affected by wind, temperature and density variations which reflected or refracted sound energy back to earth. Between 60 and 100 miles from the explosion there was no sound, although sound was heard again at greater distances. An explosion at La Courtine was heard in Paris over 230 miles away. See Richardson, E .G., *Sound*, London, 1929

[21]   In correspondence between Professor Clifton and Sir George Stokes, President of the Royal Society, Clifton expressed the view that Tyndall's observations were of great importance: '... the unsatisfactory hypotheses commonly adopted that sound can be greatly weakened, indeed practically destroyed by reflections at the surfaces of minute globules of water in the air is shown to be unsupported by the facts ... .The presence of aqueous particles in ... sufficient quantity to seriously impair ... optical transparency ... is compatible with great acoustic transparency ...' (March 1874)

*'Thus by slow degrees the caprices of the atmosphere made themselves known to us;
showing that even within the limits of a single day the air, as a vehicle of sound,
underwent most serious variations'.*

The importance of conducting the trials under different meteorological
conditions was clear:

*'... each succeeding day provided us with a virtually new atmosphere, clearly
showing that the conclusions founded on one day's observations might utterly break
down in the presence of the phenomena of another day'.*

On another fine day with poor sound transmission, Tyndall arranged to steam
towards the lighthouse on axis to find the point where the sounds could just be
heard. Subsequently venturing closer into the cliffs, the sounds became distinctly
audible but were additionally accompanied by aerial echoes which were being
strongly reflected to the observers aboard the steamer, but from no obvious visible
surface, the atmosphere being completely clear.

This phenomenon had been predicted by Tyndall as a complementary effect of
the flocculant air which had prevented the sound from penetrating out to sea. He
believed that an invisible, but impervious, 'acoustic cloud' was reflecting the sound.
He thought that the intensity of the echo was inversely proportional to the distance
that sound penetrated through the air from the source.

A gig was lowered from the *Irene* and rowed to the foot of the cliffs to listen to
the echoes from the siren:

*'The echoes returned by the transparent and perfectly invisible atmosphere were of
astonishing strength and sweetness; 'more wonderful than ever', as one of the
observers remarked at the time ... the average duration of the siren echo was 11
seconds ... October 16th, warm sun, WSW 1'.*

Throughout the trials the siren was usually operated by steam at 70 psi but tests
with lower pressure and at lower speeds of rotation were also carried out and, in
each case, the maximum audible range and the 'hardness' of the note was recorded.

The normal running speed of the siren was 2,400 revs per minute, which
produced a note of about 480 Hz.[22] It transpired that pitch variations under different
meteorological conditions did not produce markedly different results. A note of
400 Hz had improved penetration, as did a pressure of 80 psi, but the difference
between 40 psi and 80 psi was too small to justify the higher running costs.

Although steam and air were used to sound whistles and reeds in the trials,
surprisingly, compressed air was not tried with the siren, which was exclusively
sounded by steam.

---

[22]   On 18th October the siren was run at 1500 revs per minute to produce a note of 300 Hz,
approximately D sharp, then at 1800 rpm to produce 360 Hz, approximately F sharp, then at 2100 rpm
to produce 420 Hz, approximately A flat and then 2,400 revs to produce a note of 480 Hz, which is
approximately B flat.

The siren was distinctive and impressive out at sea but it was also audible ashore. On at least one occasion James Douglass visited surrounding villages near South Foreland during the trials;

> *'Proceeded to Walmer, distance 3¾ miles. Could not hear gun or syren in any part of the village. Informed by the landlady that she frequently heard the fog horns in the house, and they sometimes made her quite miserable'!*

A farmer's horse and cart standing at the mill in the village was observed responding to the sudden commencement of the signal: '... the horse took fright and ran away. Wind, E2, Cloudy'.

Neither Douglass nor Tyndall attended all of the trials, which were generally undertaken by staff from the recently-established Engineering Depot at Blackwall. Professor Holmes was present for some of the trials as was E. Price Edwards and other Trinity House staff and members of the Elder Brethren. Board of Trade members visited, as did officials from other lighthouse authorities, including the Commissioners of Northern Lighthouses and the US Lighthouse Board.

The South Foreland trials took place between 19th May, 1873 and 21st February, 1874 and over the course of the trials it appears that the guns sometimes fared better than the other signals and the steam whistle was sometimes more impressive than the reed. The performance of the siren, however, proved consistently better than all the other signals under most circumstances. Tyndall was pleased with the performance of the siren and fascinated by some of the surprising information about sound transmission, which he had not anticipated.

The trials indicated that anomalies in sound transmission produced by temperature, humidity and differences in surrounding pockets of air, could reflect sound, produce silent areas and vary the audible range of a signal from half a mile to over 12 miles. It was confirmed beyond doubt that the apparent loudness of a signal should not be interpreted as an indication of proximity to the source.[23] With these considerations in mind Tyndall's report to the Elder Brethren urged that the mariner should always err on the side of safety:

> *'The mariner when he hears a fog signal, ought to assume the minimum rather than the maximum distance ...'*

---

[23] In the early 1880s, Lieutenant Commander Chadwick of the US Navy observed that mariners should not associate the intensity of the signal with proximity, nor should the relative sound level or previous experience be used as an indication of approximate position, nor should the mariner assume the signal has ceased sounding because he fails to hear it when it should be well within ear-shot. He also said that the mariner should: 'not expect to hear a fog signal as well when the upper and lower currents of air run in different directions; that is, when his upper sails fill and his lower sails flap; or when his lower sails fill and his upper sails flap'.

  Chadwick was also very specific about the worst conditions for hearing sound: '... in the atmosphere of a clear frosty morning on which the warm sun has risen and been shining for two or three hours'! Johnson, A.B., *Aberrations of Audibility of Fog Signals*, Washington, 1885.

Fog signals could not provide a reliable means for navigation, but could provide important warnings. Tyndall recommended that Trinity House should install sirens as standard fog signals at major landfalls, although he conceded that reeds were suitable for some sites. He further recommended that a siren should be fixed clear of obstacles, such as rocks and headlands, though not necessarily at great height above sea level.[24] For optimum advantage the horn should be angled at 15 degrees below the horizon and should be designed to be easily turned to face into the prevailing wind. Against the strongest wind the siren could usually be relied upon at two miles distance in fog. Regarding the duration of the blast and the interval between, Tyndall thought that the blast should not be too brief and that the interval between siren blasts should not exceed 30 seconds.

Not all of Tyndall's recommendations were acted on. The Elder Brethren still regarded the fog gun favourably and did not agree with Tyndall on this point. In fact, further trials with guns were insisted upon by several Elder Brethren.

The fog signal trials were very thorough but some questions remained unanswered. Tyndall had been hopeful throughout the trials of carrying out experiments in fog but, unfortunately, persistent foggy conditions at South Foreland failed to materialise. Approaching London in late October:

> '... the weather ... thickened to such a promising extent, that Mr Douglass, at my request, returned promptly to Dover, only to find the weather there clear'.

With only a few light mists at South Foreland, Tyndall ended up by carrying out some experiments with a whistle and an organ pipe near The Serpentine in London. In December 1873 Tyndall made repeated visits to Hyde Park during which he noted the variations in the range of different sounds. In thick fog, above the sound of distant train whistles and the traffic on the Bayswater and Knightsbridge Roads, he distinctly heard Big Ben striking about two miles away, but on clear days it was impossible to hear Big Ben.

The observations in Hyde Park confirmed one important result of the trials, which was that the optimum conditions for sound transmission occurred in homogenous air, as occurred in foggy weather. Tyndall was not sure if all fogs would consist of homogenous air, however. He thought it was possible that there could be temperature variations within a fog bank, and it was certain that there would be temperature and humidity differences in patchy fog. The phenomenon of partial reflection of sound and silent areas had been recorded in conditions of clear visibility in the South Foreland trials, but if they occurred in fog the consequences could be very serious. Could 'acoustic clouds' occur in some kinds of fog?

---

[24] The testing of signals at the bottom of the South Foreland cliffs was discontinued quite early in the trials when it was established that the height of the signal made little difference to its range.

John Tyndall succeeded in reproducing the effects of 'aerial echoes' and the 'acoustic cloud' in his laboratory and he subsequently wrote several papers on sound, which number amongst the Philosophical Transactions of the Royal Society.

Peace and quiet was restored to South Foreland on completion of the trials when all the fog signal machinery was removed, though the nearby South Sand Head lightvessel[25] would shortly be equipped with one of the early sirens.

Though Tyndall's report was less conclusive than the Elder Brethren originally hoped for, it shed new light on many of the questions posed at the start of the fog signal trials and had strong bearing on the development of the siren and related equipment subsequently manufactured in the Engineer's Department at Blackwall workshops.

---

[25] South Goodwin lightvessel.

Chapter 4

The Lizard

and the development of the siren

The House of Commons Committee which reported on fog signals in 1873 recommended a reed for the Lizard, but events overtook the implementation of the proposal. Although the Elder Brethren were convinced of the merits of a Holmes' reed, construction of a fog signal at the Lizard was postponed because of the South Foreland trials. It was not until 1878 that a fog signal came into regular service at the Lizard, but it was not a reed horn after all.

The Lizard Lighthouse.

Before the South Foreland tests, James Douglass had drawn up plans for a fog signal gun at the Lizard. The drawings, dated 1873, indicate that a purpose-built fog gun house with a separate magazine was intended for the cliff edge a little way to the south and west of the lighthouse. Eighteen-pound muzzle-loading cannon were to have been used, but the Lizard battery was not built and the fog guns were never installed.

The Admiralty List of Lights first refers to a fog signal at the Lizard in the 1876 edition. The lighthouse:

> *'will shortly be altered to electric … and a fog horn will also be established'.*

When the 1878 edition was printed, the Hydrographic Office could only state that the:

> *'Fog Trumpet* will be *one blast every five minutes'.*

The original Lizard light was built in 1619 but the two existing towers were established in 1751 by Thomas Fonnerau. The twin towers were originally lit by coal brazier and kept bright by bellows blowers. The lighthouse superintendent lived in a small cottage between the two towers and, should the fires dwindle, he apparently encouraged the bellows blowers into action by blowing on a large cow's horn![1] There are no references as to the application or possible effectiveness of this instrument as a fog signal!

When Robert Stevenson visited the Lizard lighthouse in 1801 both towers were still lit by coal, but oil lamps were in service by 1813.

The 1859 Bristol Channel Pilot reminded its readers that the Lizard:

> *'… cannot fail to be known by its two lighthouses which exhibit fixed lights from Argand lamps and reflectors …'*[2]

Michael Faraday was the Scientific Advisor to Trinity House during the 1850s. From 1839 to 1855 Professor Faraday reported on a wide range of discoveries about electricity, particularly electromagnetism, in his *Experimental Research on Electricity*. Frederick Holmes, who had been living and working in France, returned to England in 1856 with plans for developing a magneto electric generator. Holmes patented his idea and made proposals to Trinity House in 1857. Faraday and Holmes experimented with magnetos near Blackwall and, although the output was only 1½ kilowatts, the Elder Brethren approved a full scale experiment at the South Foreland lighthouse. These magnetos were 8 feet in diameter with the coils rotating about a fixed magnet. Electric light was first shown at South Foreland lighthouse in December 1858 and a similar magneto was subsequently installed at Dungeness where electric arc light was in use from 1862.

---

[1]   *Journal of the Royal Institute of Cornwall*, 1871–1881.

[2]   Argand lamps (named after the Swiss inventor) were introduced from 1790. The circular wick and glass chimney consumed more oxygen and greatly increased light output.

The 'improved' magneto design, which Holmes demonstrated at the Paris Exhibition in 1867, was used at Souter Point lighthouse. This machinery ran at 400 revs per minute and was a great improvement on that installed at Dungeness. A 5 bhp Whitworth 'Allen' steam engine drove the magneto at Souter Point.[3]

Throughout this period, comparative trials of equipment continued with magneto designs and the developing dynamo electric machines. Different types of carbon rods[4] used to create the arc light were also tested. At Dungeness, deposits of carbon, which had dropped from the rods, had damaged the lenses and impaired the overall efficiency of the light. At both South Foreland and Dungeness the results of electrification were disappointing as the brightness of the light was much less than expected. Operating difficulties resulted in the standby oil lamps being resorted to and James Douglass calculated that it cost two or three times more than oil to operate the arc lights.[5]

The electrical equipment at Dungeness was all replaced by oil lamps but, despite these setbacks, Trinity House proceeded with new trials within a year or two. Dr Tyndall continued the electrical experiments when he succeeded Faraday in 1866. As the Scientific Advisor to Trinity House, a considerable amount of his time was spent on electrical research, which continued into the 1880s.

Dr John Hopkinson took over the running of Chance Brothers, the lighthouse manufacturing company, from James Chance in 1872 and was responsible for many developments in the design of lighthouse optics and engineering. Hopkinson was familiar with the work of Holmes and Faraday and developed his own work on dynamo design when Chance Brothers became involved in the experiments with electricity at South Foreland in the early 1870s. Hopkinson wrote various articles on the configuration of windings and the laws of magnetic circuits. On leaving Chance Brothers, Hopkinson remained as consultant to the company and he also became consultant to the Siemens company. He retained patents for his dynamo designs and when Dr Tyndall and James Douglass recommended the Siemens dynamo for the Lizard lighthouse, Hopkinson designed the electrical equipment and the optical equipment installed in both towers.

The Siemens' D2 dynamos, installed in 1877, were horizontal machines which produced 40 amp current of 50 volts when driven at 850 rpm. There were six of

---

[3] The original 3¼ hp engine was replaced by the 5 hp engine in 1871. In 1878 a 20 hp steam engine was installed.

[4] The carbon rods used at the Lizard were 1½ inches in diameter and burned about 2 inches each hour.

[5] At South Foreland in 1884, further tests were made into the relative merits of oil, gas and electricity for lighthouse illumination and debate continued about the cost and effectiveness of whale oil, mineral oil and gas, beside electric light. Douglass considered oil to be the most efficient. Although Souter Point, the Lizard and St Catherine's were not converted back to oil, there was no further development in electrification until the 1920s when The Dover Electric Supply Company linked South Foreland to mains electricity (1922). A filament lamp was first used near New York by the US Lighthouse Board about 1900 and introduced at English lighthouses from the 1920s.

these machines bolted to a common bed-plate and driven by a leather belt from an overhead shaft. The cost of the electrification and the new engine house was estimated at £15,000, but the total cost reached £22,500, when the work was completed in early 1878.[6]

The engines used for driving the magneto machines at Souter Point and St Catherine's Point used a common drive shaft to power the air compressor for the reed horn fog signal and, as at South Foreland and Dungeness, steam engines were used to provide the power. At Souter Point lighthouse the staff consisted of one engineer and four attendant keepers, who had all been trained in the new skills at Dungeness or South Foreland electric lighthouses. The role of the engineer was supervisory but the attendant keepers operated all of the equipment and were required to keep watch in the lantern room and in the engine room when the light was in operation. This new operating practice was aided by communication between both rooms using copper speaking tubes similar to those used aboard ships. Similar speaking tubes still exist at the Lizard, connecting the lantern room and engine room with the keepers' quarters. Five keepers were also on duty at the Lizard.

There was some discussion about the staffing levels of the electric lighthouses concerning the number of staff and the new duties required of them. Some people felt that a lightkeeper's main duties were compromised by requiring him to supervise boilers. Others felt that keepers might require pay rates commensurate with firemen and engineers. There were also safety considerations regarding the use of the steam engine, both for the lighthouse and for the keepers. With the increased risk of fire there was also the additional danger of a boiler explosion, a regular cause of industrial accidents at the time.[7] In America at least one lighthouse was destroyed when the fire for the steam fog whistle ignited the fuel store.

Steam-powered whistles and sirens were not widely used as fog signals in Britain and steam engines were generally confined to the electric lighthouses, where the operation of the fog signal was a supplemental duty to generating electricity.[8] Steam engines were not suitable for compressing air for fog signals as starting a steam engine from cold was a lengthy process and burning fuel on standby would

[6] *Parliamentary Proceedings* 1880. Maintenance costs for the Lizard lighthouse in 1879 were £2305.

[7] Boiler explosions were common, often attributable to the attendants or 'undue pressure, shortness of water...' *Engineering*, October 1871. There were over 60 boiler explosions in 1871 which killed 36 people and injured 113. Fourteen of the dead were 'strangers or passers-by'. At ironworks, mills, mines and elsewhere, corrosion, weak tubes and poor repair rendered steam boilers very dangerous. The construction of some boilers was such that they were barely adequate for the pressure sustained in normal use. The Board of Trade finally undertook investigations into accidents under the Boiler Explosions Act of 1882.

[8] The electric lighthouses at Dungeness, South Foreland, St Catherine's, Souter Point and later at the Isle of May in Scotland, all used steam engines to generate electricity. John Bowen's 1947 book *British Lighthouses* implies that the Lizard also had steam engines but an accompanying illustration shows one of the Brown engines installed by Douglass.

have been extravagant. The adequate provision of coal and water at isolated stations was likely to be difficult and for reasons of safety and efficiency the use of salt water in boilers was not recommended. These considerations mitigated against the general use of steam engines and steam fog signals, although the Brown siren, used to such good effect during the South Foreland trials was, of course, steam-operated!

A. & F. Brown also supplied hot air engines to the South Foreland trials and the simplicity and economic operation of these engines impressed Trinity House. James Douglass, the Engineer in Chief, was not enthusiastic about steam engines. Hot air engines were less powerful than high pressure steam engines, but they were practical and economical.

Small hot air engines had been used aboard lightvessels for a few years and after due consideration, Douglass recommended that hot air engines should be used at the Lizard, both for the fog signal and to drive the electric dynamos. Three 10-horse power caloric engines were consequently ordered from A. & F. Brown of New York and installed in the new engine house.

A workshop, storeroom and coke store were incorporated in the engine house which was built in front of the two towers and linked to the main buildings by a stepped corridor. Extra accommodation was built at the same time for the increase in staff.

> The Minutes of the Proceedings of the Institution of Civil Engineers contain an article written by James Douglass in 1878 in which he describes the equipment he installed at the Lizard. Each of the three 10 hp engines had a single-acting vertical cylinder. Standing adjacent to the cylinder on each bed-plate was a large brick-lined iron furnace and integral coke feeding hopper. The 25-inch diameter piston had a 20-inch stroke and rocked a beam which drove a large flywheel at 60 rpm. The flywheel, 8 foot 6 inches in diameter, produced the down stroke and also drove a small air pump and valve gear. The air pump forced air through the coke furnace which burned fiercely and expanded the air to a pressure of 26 psi. The hot compressed air then escaped through the main cylinder pushing up the rocker beam and driving the flywheel.
>
> The air compressor for the fog signal was attached to the engine bed-plate and had a single acting 9-inch diameter piston with a 20-inch stroke capable of compressing air to 50 psi.

The huge flywheels on the caloric engine used belts to drive the overhead shaft at the front of the engine house. Behind and below the shaft were the six Siemens dynamos arranged in pairs before each of the three engines. The drive belts could be slipped on and off the flywheels as necessary and in most conditions only one

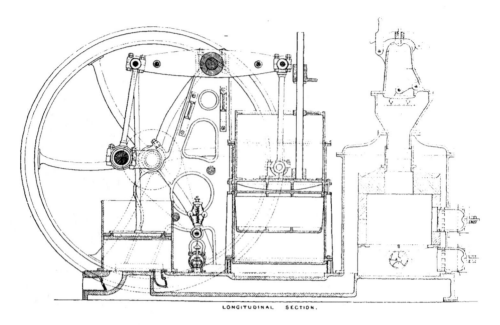

LONGITUDINAL   SECTION.

The caloric engines installed at the Lizard. Courtesy Institution of Civil Engineers.

engine was required to power the two lights.[9] During fog, one engine could be used exclusively for the fog signal by disconnecting two of the dynamos.

During the day a very low fire was kept in one of the engines so that if fog set in the fire could be drawn quickly and brought up to temperature.

The keepers at the Lizard were visited by Trinity House engineers from the *Galatea* on 13th June 1879 when the fog signal was tested:

> 'Fog horn fire banked up and in 7 minutes had a blast guage up to 17 and down to
> 15. The second Blast guage up to 25 and down to 21½. Third Blast guage up to
> 32 and down to 27. Applied a Match to the Coal fire and had a Blast in 29
> minutes. Guage up to 42 and down to 38'. (sic)

It is clear from this entry in the station Order Book that the signal could be sounded at considerably lower pressure than the nominal 50 psi intended. Between each of the 5 second blasts of the fog signal, the engines had nearly 5 minutes to pump up the pressure in the two adjacent air receivers.

The fog signal installed at the Lizard in 1877 was a siren but although Douglass had bought the Brown caloric engines, a Brown siren was not purchased. The siren

---

[9]   Although Douglass calculated the cost of electric light at Dungeness as being three times more expensive than oil, he was satisfied with the improved facilities at the Lizard. He thought the improvement in the engine, dynamo and arc lamp produced 20 times the quantity of light produced at Dungeness for about a tenth of the cost!

Inside the engine house at the Lizard in the early 1880s. Courtesy of Trinity House.

installed was designed by James Douglass and George Slight, who ran the engineering workshop at Blackwall. The American siren was a disc but the new instrument at the Lizard was of a cylindrical design 6 inches in diameter and 9½ inches long. It had 12 half-inch slots, 8¼ inches long, cut in the cylinder walls and there was a single slot half an inch by 8¼ inches long, which formed the throat of the trumpet where it was attached to the siren chest.

Like the American siren, the moving part of the instrument, the rotor, was driven up to speed mechanically before air was released through it from the receivers. The caloric engine also regulated the character of the signal – one 5 second blast every 5 minutes. The siren produced a similar note to the American disc, about 480 Hz.

The Lizard siren entered service on a foggy Sunday in January 1878. It appears to have been no less an instrument than the Brown siren and was equal to '... if not exceeding, in the discordancy of its note, Dungeness'.[10] A local reporter thought that the sound was

> '... not so loud and disturbing as was anticipated (but) the sound is very weird and melancholy ... (and) ... there it rolls, with prolonged reverberating echoes through the surrounding precipices and caves'.[11]

---

[10] Hardy, W.J. *Lighthouses, their History and Romance,* 1895.

[11] *The Cornish Telegraph,* 22nd January 1878.

The siren had a single cast iron trumpet which was 15 feet long and opened out to a diameter of 18 inches at the mouth. Early photographs show the horn lying horizontally on the flat roof at the front of the engine house. The horn was designed to be turned to face into the prevailing wind, as recommended by Tyndall in his report on the fog signal trials, and there was a small wheeled carriage near its mouth for that purpose. The length of the horn slightly exceeded the radius of the curved bay at the front of the engine house and protruded about a foot over the iron guide rail and the wall of the building.

Similar instruments were installed in 1878 at Start Point and Bardsey and in the previous year, Trinity House had installed sirens at St Ann's, the Skerries and Casquets, all with horizontal horns on top of predominantly circular buildings.[12] By the end of 1878, six lightvessels also had sirens. In 1884, ten years after the fog signal trials at South Foreland, there were

'... *twenty-two sirens* ... *on our coasts and sixteen aboard lightships* ...'.[13]

The South Foreland siren trials used different steam pressure and varied the speed of the disc rotation between 1500 and 2400 rpm. Exhaustive testing subsequently indicated many of the shortcomings of the instrument. The work of George Slight at Blackwall Depot improved on the original design and construction by experimenting with the number, size and shape of the ports cut into the rotor and stator of the siren. Slight's main contribution at an early stage was the idea of abandoning the disc in favour of a cylinder, which was mechanically more efficient and capable of uniform running at slower speeds. However, the Lizard siren rotor still ran at 2400 rpm.

Frederick Holmes must have been a regular visitor to South Foreland in connection with the magneto and dynamo experiments and, as previously noted, he was also involved in the earlier development of the reed horn – he had patented his own designs in the early 1860s.

Reed horns designed by Holmes were in use on several light vessels and Holmes' reeds were also tested in the South Foreland trials. George Slight's engineering developments were part and parcel of his duties at the Trinity House workshops, but Frederick Holmes had an eye for business and an interest in filing patents. Holmes was once credited with developing the cylindrical siren and with introducing the 'automatic siren', although the principle was familiar from Tyndall's laboratory model. In any case, Holmes patented the automatic siren in 1875.

In the automatic siren, instead of the rotor being mechanically driven up to speed before the air was admitted, the cylinder was driven round by the escaping

---

[12]   The typical cost of fog signal equipment in 1879 was £1800 for the building and £2050 for the machinery, totalling £3850. The annual running costs (based on 360 hours of fog) totalled £334 1s. 4d., which included interest on first costs plus stores, fuel, wages and repairs (£109). Robin Allen, Trinity House Secretary, Parliamentary Proceedings, 1880.

[13]   Price Edwards, E. *Seamarks*, 1884.

The original fog trumpet and two electric lanterns, circa 1890. Courtesy Helston Folk Museum.

Detail of the trumpet and carriage. Courtesy Helston Folk Museum.

air itself. By cutting the slots in the cylinder walls at opposing angles in the stator and rotor air impinged on the bevelled surfaces along the edges of the slots, causing the rotor to move and then to rapidly accelerate. The characteristic sound this produced was a rising 'whoop' at the beginning of the blast. On steam ships and later on steam traction engines, the automatic sirens produced a rising note throughout the brief duration of the blast, but for use as a fog signal, a sustained note of a certain pitch was required from the siren for most of the duration of the blast. Braking devices were introduced to prevent the rotor exceeding the set speed.

These governors were cork-tipped weights on the siren shaft which were thrown out by centrifugal force against the cylinder wall to slow the rotation. Slight perfected his 'improved' governors and standardized the whole instrument by 1880. This was the 5-inch diameter siren with 30 ports.

Other experimental cylinder sizes of 4, 5, 6 and 8 inches in diameter were made for James Douglass as engineering drawings held by Trinity House indicate. Sirens were made by several ship builders and brass foundries and many steam ships, including those in the Royal Navy, were fitted with sirens. Sydney Smith of Nottingham made a range of sirens for different purposes, including signalling for military use, factory use and for use on steam engines and ships.

In France, Sautter Lemonnier produced an automatic brass siren with copper trumpets which could be sounded by air or steam at about 70 psi, and these were used on the ships of the Compagnie Transatlantique. These sirens could be used for fog signalling but also had a manual control enabling them to be used for ship-to-ship Morse code communication. Other French sirens introduced by the lighthouse authority were as small as 3 inches in diameter but were normally about 6 inches. Sautter Lemonnier also introduced a double-noted siren in 1880 which had two sirens which simultaneously sounded in the same trumpet to produce a distinctive sound and an increased range for the instrument.

The proliferation of sirens ashore and afloat began to pose difficulties[14] and mariners argued that the similarity of signals was misleading: it was impossible to tell if a sound came from a lighthouse or a ship. In recognition of the problem the Board of Trade and the Elder Brethren of Trinity House consequently invited Professor Holmes to design an improved instrument which would produce a much more distinctive sound for lighthouse fog sirens. Holmes came up with an effective solution in 1881 and in 1882 he patented the two-note siren. Unlike the French example where the two sirens sounded together, this instrument sounded two different blasts, a high note and a low note.

The high and low note instrument enabled a highly distinctive characteristic to be sounded – by combining three blasts different lighthouses could sound low, low, high; high, low, high; low, high, low and so on.

The Holmes' siren used a redesigned braking system. Instead of centrifugal governors determining the maximum speed and hence the pitch of the notes, Holmes' siren had two discs attached to the shaft at the base of the siren. Pressure on the edge of the disc regulated the speed of rotation and by controlling the braking on each disc, two constant but distinctive notes could be produced.[15] Professor

---

[14] Steam vessels were required to sound one blast of the whistle or siren if on the starboard tack and two blasts if on the port tack.

[15] Two air-operated pistons impinged on the edge of the upper disc to prevent the note rising above the intended high note, and after the appropriate interval a second pair of pistons acted on the lower disc to produce the low note. Both sets of pistons were fed by the same air supply as the siren so that as the pressure dropped during the blast, it dropped equally to the piston to maintain the same note. Holmes' design was granted patent number 3528 in April 1882.

Holmes had his own manufacturing company[16] which produced the equipment complete with hot air engines for sale in Britain and abroad.

An alternative two-note design was developed and used in Trinity House sirens and Holmes' company also later adopted the same system. In this siren, two sets of ports were cut into the same cylinder, one set above the other. The high note was produced by 32 ports and the low note by 14 larger ports. Two separate air valves supplied air to the high and low note ports and the same centrifugal governors controlled the maximum speed of revolution of the siren as it produced the separate successive notes. The intended frequency was 670 Hz for the high note and 290 Hz for the low note.

The two-note sirens were steadily phased into service at lighthouses and lightvessels and the earlier means of producing a distinctive character using multiple blasts and varying length blasts were superseded. Most of the new two-note sirens gave three blasts.

The automatic operation of these signals was controlled by actuating machinery which was originally directly driven by the fog signal engine, as at the Lizard, but other means of operation were subsequently introduced which used compressed air. The air drove a geared cam which had notches corresponding to the character of the signal and when the notches tripped a weighted lever attached to a valve, air was released through the siren. Holmes' sirens generally used this type of device.

An improvement on the weighted lever was the introduction of the pneumatically-operated admission valve which was installed between the siren and the receiver. The pressure of air in the system kept the admission valve closed, except when controlled pressure on the rear of the valve shaft opened it to allow air through to the siren.

> A clockwork actuating machine was devised to automatically produce the required characteristic. A small bore air pipe was led from the receiver to the actuating machine and from there to the admission valve. A spring-loaded poppet actuating valve in an elbow joint stopped the air flow in the pipe until a cam on the actuating clock released it. Different sized cam wheels produced longer or shorter periods of silence, and the duration and number of blasts was produced by corresponding notches cut into the circumference of the cam wheel. The notches released pressure on the actuating valve which in turn opened the admission valve and sounded the siren.[17]

[16] The Caloric Engine & Siren Fog Signal Company Limited had their works at 24 Budge Row, Cannon Street, London EC.

[17] An 1882 drawing of a 'Machine for actuating siren admission valves' shows gearing ratios and cam sizes for sirens at Bardsey, Casquets and Start Point lighthouses. The cams for Start Lighthouse took 180 seconds to revolve and regulated the character as follows: high 2 seconds, silent 2 seconds, low 2 seconds, silent 2 seconds, high 2 seconds, silent 170 seconds. Bardsey and Casquets both had identical characteristics at that time, 3 short blasts in 5 minutes; blast, 2 seconds, silent 2 seconds, blast 2 seconds, silent 2 seconds, blast 2 seconds, 290 seconds.

The original Lizard siren was altered in 1889. In July 1888 the Order Book shows that the siren was 'keeping time', but in December there is a reference to a change of signal and the Admiralty List for 1890 described the new signal as having two quick blasts, high and low, every 2 minutes.

> The new 6-inch siren had high and low ports and was designed to work at 25 psi. There were 24 ports, 3.45 inches long and 0.29 inches wide to produce the high note and the low note was produced by 16 ports of 2.25 inches in length and 0.5 inches in width. The driving area on both sets of ports was 20.7 inches and the cylinder ran at the same speed to produce both low and high notes.

At the Lizard, the 6-inch two-note siren replaced the original single note 6-inch siren, possibly in the same chest, but obviously with alterations for supplying the high and low note ports. As no other major changes were needed to ancillary equipment, the siren remained in the same position at the front of the engine house with the two original air receivers just to the rear. There would have been ample air, as the blasts were shorter, despite being more frequent. The siren retained the original 15-foot horizontal horn on the roof at this stage and the original Brown caloric engines from 1877 continued to provide the power.

There had been other alterations in the engine room by this time, however; the original Siemens dynamos had been replaced by two new machines.

The *Engineering* journal of October 7th 1881 illustrated the 'De Meritens Magneto Electric Machine shown at the recent Paris Electrical Exhibition'. In principle, the De Meritens machine returned to the Holmes magneto design, but was much more efficient. Each magneto, which weighed 4½ tons, could deliver 4½ kilowatts of alternating current at 830 rpm. Trinity House bought two machines for St Catherine's lighthouse and two for the Lizard, where they came into service in July 1885.[18]

Both machines remained in service at the Lizard until June 1950 when mains electricity finally made them redundant.[19]

Thomas Matthews succeeded James Douglass to the post of Engineer in Chief to Trinity House in 1892 and a drawing signed and dated by Matthews in 1893[20] shows various changes to the floor plan of the Lizard engine house. On the east side of the engine house were workshops for the 'smiths' and 'carpenter' and on the west side, the oil store and coke store. Behind the two air receivers at the front

[18]   A plaque on the surviving magneto says it was installed in 1881.

[19]   The arc lamp and carbon rod was replaced by a 3 kilowatt electric filament lamp in 1926. Between 1926 and 1950 the magnetos provided the power for the filament lamp via a transformer.

[20]   Trinity House engineer's drawing No. 7026.

of the building were the De Meritens magnetos connected by a common overhead drive shaft to four proposed new oil engines.

> Otto's four-stroke cycle had been patented with the introduction of his horizontal gas engine: as the piston travelled towards the crank shaft the fuel mixture was drawn into the cylinder, and on the return stroke this mixture was compressed to the point at which ignition occurred near top dead centre. The 3rd stroke was the explosive travel of the piston down the cylinder and on the 4th stroke the exhaust gases were vented. Thousands of the German engines were made and exported all over the world and the engines were also made under licence in some countries. The Crossley Brothers built engines under Otto's patent in Manchester as early as 1875.
>
> The Priestman Brothers of Hull were originally boat builders who produced a French petrol engine under licence. They experimented with heavier oils and developed their own oil engine which they patented in 1886. The Priestman oil engine incorporated an air compressor to spray the oil into a vaporizer. This mixture was heated by the exhaust gases and the hot mixture was then drawn, with air, into the combustion chamber and fired by a spark plug. Horizontal and vertical types up to 100 hp were made, generally for the agricultural market but, interestingly, they also found a use in '…lighthouses for driving air compressors to work fog signalling apparatus'.[21]

The Lizard Order Book first refers to the 'new oil engine' in June 1894.[22] For one reason or another, only one of the four oil engines proposed by Thomas Matthews was actually installed at the Lizard and the caloric engines remained in service to drive the magnetos until 1908. A visitor to the lighthouse described the engines as 'three very antiquated hot air affairs and one … oil engine'.[23] The Priestman engine was situated at the west end of the engine house.

The 'antiquated' Brown caloric engines were dismantled in 1908 and replaced by three 14 brake horse-power Hornsby oil engines.[24]

The 'hot bulb' engine designed by Herbert Ackroyd Stuart was first produced by Richard Hornsby & Sons in 1891. Single cylinder Hornsbys were widely used by Trinity House at lighthouses and aboard lightvessels and a more detailed description of these engines can be found in Chapter 10. The three Hornsby engines at the Lizard were later joined by a fourth when the Priestman was removed in June 1921

---

[21] Edginton and Hudson, *Stationary Engines for the Enthusiast*, 1981.

[22] A Trinity House engineer's drawing from 1894 shows a Priestman engine with an air compressor directly coupled to the crank shaft. The compressor was made to Trinity House specifications by Johnson & Hutchinson of Stratford, East London.

[23] Stone, John Harris, *England's Riviera*, circa 1910.

[24] The three original Hornsby engine serial numbers were 30338, 30339 and 30340. The number of the later Hornsby is not known.

Engine room interior in the 1920s with Hornsby oil engines. Courtesy Trinity House National Lighthouse Centre.

and these machines remained in daily use with the De Meritens magnetos until 1950. With the arrival of mains electricity, the magnetos were disconnected.

The Hornsbys were replaced by four new engines in 1950, three for use with the fog signal and one for use as a generator for the light in the event of mains failure. The Gardner diesel engines which replaced the Hornsbys were widely used by Trinity House and those at the Lizard were in use for nearly 50 years.

The 4 cylinder Gardner which runs at 1000 rpm is much noisier than the Hornsby and at 40 bhp it is considerably more powerful. The smell of the warm engine, lubricating oil, diesel, grease and metal polish is part of the fabric of the working engine house and since the introduction of high speed diesel engines, so is the noise. The sound of the Hornsby quietly and slowly compressing the air may have been quite delightful but the sound of the Gardner can be exciting!

With the engine running and the stop valves open to the receivers, the clutch is wound in and the Vee-belt drive engages with the Reavell compressor. The noise level increases further. The rotary compressor has 4 × 7½ inch diameter double-acting pistons with a 5-inch stroke. It is started off-load and then the two unloading valves are wound in to pump air to the receivers, at which point the noise level falls

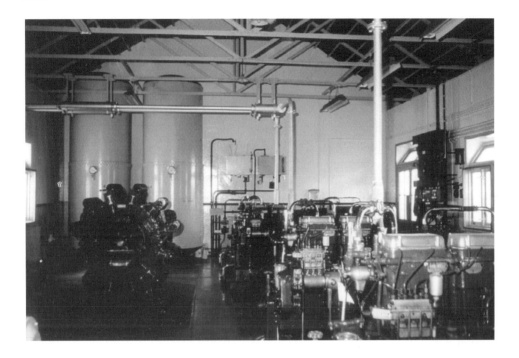

The engine room today with the magneto on the left.

a little. Once the air main from the receivers is opened to the sounding receiver below the siren, the coders can be switched on and the siren starts.

The Lizard engine house has not changed very much over the years, the Gardners now stand where the Hornsbys used to, one of the magneto machines remains on site and the siren equipment still stands in the bay at the front of the building. The slate floor shows signs of other bolt holes which have been restored over the years. Excepting the magneto, there is no other trace of any original or early electrical equipment. However, the control panel still in use on the front wall is contemporary with the engines. It is similar to equipment installed aboard post-war lightvessels. The impressive dials, scissor switches, Bakelite and brass mediate the current generated by the engines which supplies the coder for the fog signal. Each of the three Gardners drives a 2-kilowatt, 100-volt DC motor, but only one of the engines is normally required to operate the fog signal. The fourth Gardner is coupled to two generators and would have powered the lighthouse in the event of mains failure.

Entering the engine room from the main buildings gives a view directly ahead of the siren at the front of the building. On the right are two engine and compressor sets, the antique magneto and the air receivers. On the left is a compressor set and disused generator set. On the right of the siren bay is a semi-diesel Blackstone engine but this was never part of the Lizard equipment.[25]

Compressed air is stored in the receivers at 50 psi and the siren can therefore be started immediately when required. Storing air has been the general practice at most lighthouses since 1886, although the stored air pressure was much higher at that time, up to 125 psi.[26] The advantage of storing air in these 'batteries' was that the siren could be operated whilst the oil engines were being started up: it could take up to 20 minutes for the Hornsby engine or Brown engine to begin compressing air.

As the siren was operated at 25 psi, a reducing valve was installed between the smaller sounding receiver and the high pressure main storage receivers. When the oil engines were operating normally they would still be pumping at a slightly higher pressure than 25 psi and this maintained a higher pressure in the main receivers.

Once the fog had cleared and the siren was stopped, the engines ran on until the receivers were filled to the required pressure, up to 'six or eight atmospheres'.

Before storing high pressure air became regular practice, the compressor, air receiver and siren were all open to the same air pressure, originally about 50 psi, but this figure was the maximum pressure in the system and the siren would have sounded at half pressure without problems. The two original receivers at the Lizard were located beside the siren at the rear of the bay and at the front of the engine house. These receivers were scrapped during the major refurbishment of the fog signal in 1908. The existing receivers at the Lizard are typical of those used generally at Trinity House fog stations and are made of rolled half-inch thick iron plate with double and treble riveted butted seams.[27]

The receivers are linked by large bore underfloor pipes to the compressors and connected to the siren via a stop valve and reducing valve with a Vernier gauge set at 25 psi. The fog signal was usually tested once a week and the engines occasionally started between times to be sure that the receivers were topped up. Periodically the receivers were emptied for inspection and cleaning.

Compressed air contains oil and water and a bilge could build up inside the receivers. Engineers from Penzance depot entered the receiver via a manhole and climbed inside to check for evidence of pitting and rusting to the walls and rivets.

---

[25] This engine was used at St. Ann's Head until the siren was scrapped in 1972. Blackstones were used at Casquets, Longstone and South Stack but did not enjoy the wide patronage Trinity House afforded Gardners of Manchester, Reavell of Ipswich and Hornsby of Grantham. One of the 5 hp Hornsby engines from Sark now in the Lighthouse Museum in Penzance was temporarily housed at the Lizard.

[26] Compressed air technology was quite familiar by this time: the pneumatic drill was invented by Ingersoll in 1870 and compressed air was used in the mining and railway industries. At the turn of the century even clocks in some New York offices could be operated by compressed air!

[27] It is not known if the receivers were made by local ironworks, by Blackwall Depot or by another contractor. Smaller receivers sometimes had overlapping seams, later receivers were more usually made of steel with welded seams.

The Gardner engine and Reavell compressor.

The tanks were periodically coated inside with red oxide paint. Accumulated water and dirt was regularly blown out of drain cocks on the bottom of the receivers along with particles of rust and dirt which, if conveyed through the system, could otherwise jam valves or even damage the siren.

Regular maintenance has preserved these receivers in good condition and although the safe working pressure is never exceeded the tanks are capable of much higher pressure.[28] The receivers stand 14 foot 6 inches tall with a diameter of 5 feet and a total capacity of 540 cubic feet at 125 psi. It could take two engines some time to refill the receivers after an inspection. Each receiver has a large brass pressure gauge[29] and the pressure gauge on the sounding receiver below the siren can be seen fluctuating during each blast, as the reducing valve replenishes the air from the main receivers.

The sounding receiver has a capacity of 48 cubic feet of air. It is 4 feet in diameter and stands 4 foot high and forms the base of the siren pedestal. It contains air at the operating pressure of 25 psi to sound the siren. A Matthews drawing dated 1896 shows details of the receiver and pedestal still in use at the Lizard and similar equipment was installed at St Ann's, St Catherine's, Lundy, Pendeen, Start Point, Dungeness and Bardsey over a period of about ten years. Another drawing signed

---

[28] Safety valves fitted to the receivers were designed to blow before dangerously high pressures could accumulate.

[29] Sydney Smith of Nottingham patented the pressure gauge in 1847 for use on steam boilers.

Siren pedestal and sounding receiver.

by Matthews shows the Lizard engine house with the two original 1877 receivers still located beside the siren in the bay, but with the new Matthews-designed sounding receiver, complete with a twin two-note siren and twin admission valves. This drawing suggests that the equipment still in use at the Lizard today could have been installed prior to the removal of the original receivers and the major refurbishment of the fog signal in 1908. As seen in relation to the Priestman engines, however, the drawings themselves do not establish that such proposals were actually implemented and other evidence suggests that the single two-note 6-inch siren from 1889 remained in use until 1908, when the current twin 5-inch siren was installed.[30]

After the new series of fog signal trials at St Catherine's lighthouse on the Isle of Wight in 1901, the merits of high and low note sirens were reconsidered, and when the new fog signal was installed at the Lizard in 1908 it was a single note siren. The new twin 5-inch siren came into service on October 8th, sounding the same character

---

[30] The Order Book contains no reference to the installation of new equipment before 1908 and a photograph taken in 1903 shows that the single horn attached to the two-note, 6-inch siren was still in use at the Lizard at that time. In 1907 the fog signal character was still: high note, 2½ seconds; silent, 2½ seconds; low note 2½ seconds; silent, 112½ seconds.

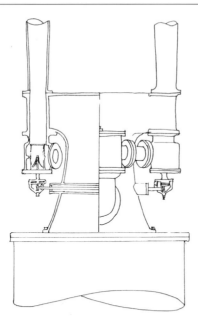

On top of the sounding receiver the cast iron pedestal supports the trumpets and siren assembly. The left side sectional view shows the siren in the chest at the bottom of the trumpet. At the bottom of the siren are governors and the gearing on the connecting rod. The right side view shows the main admission valve connected to the siren chest.

as today, a long blast of 7 seconds, silent for 2 seconds, a short blast for 2 seconds and silent for 49 seconds.

Early versions of Thomas Matthews' pedestal and siren assembly were designed for use with two-note sirens, but from its original installation at the Lizard the high-note siren ports were plated over and covers on the siren chests were tightly bolted with only the single admission valve supplying air to the low note port areas of both sirens. The shape of the siren chests indicate where ducted pipes could supply air to the high note portion of the siren cylinders. The siren chests on the flanged pedestal extend from eye level for a few inches into the 5-inch diameter throat of the two cast iron trumpets. The weight of the trumpets is supported by the pedestal which also supports the main admission valve and siren assembly. The pedestal is formed of cast iron brackets and ⅝ inch iron plate.

The trumpets extend out through the roof of the building to culminate in copper bell mouths which have a maximum diameter of 6 feet. The overall length of each trumpet is 22 feet. The design is typical of those installed around the turn of century and their effectiveness was assessed in the St Catherine's trials. These trumpets were not designed to be moved to face into the wind like the earlier design and stout roof stays hold them in place. Two trumpets, with a separate siren in the

A spare siren in its box. *Detail of 5″ siren showing sounding ports.*

throat of each, were more effective at covering a wide sea area but it was necessary to devise a means of synchronising both instruments for optimum performance.

At the Lizard, the sirens are synchronized mechanically by a connecting rod which is linked to the shaft of each siren cylinder by means of helical gearing. When air is admitted both sirens automatically spin up to speed and the con rod and the governors in each siren maintain an even note in both trumpets. Each siren is made of gun metal[31] and has a steel shaft with hardened bearings. The overall length is 15 inches including the gear and governors, the cylinder itself is 7 ¼ inches long by 5 inches in diameter. In case of sticking or damage of any sort, a spare siren was kept at each lighthouse, which could be fitted by the keepers.[32]

The twin sirens at the Lizard have 16 angled ports cut into the cylinder walls 1 ½ inches long and 0.175 inches wide. An additional pair of smaller 'starting ports' on opposite sides of the cylinder help to start the siren spinning.

When air under pressure enters the siren chest it forces its way through the ports in the stator by moving the inner rotor. This process spins the rotor and the coinciding

---

[31]   Gun metal alloy is made of 87% copper, 9% tin, 2% zinc, 2% lead. Alternatively if lead was omitted the quantity of copper was 86%, 12% tin and 2% zinc.

[32]   One of the spare 5-inch sirens I have seen at another lighthouse not only had the 16 'low note' ports, but also had the 24 ports cut in the top part of the rotor for the high note. The stator, however, only had the 16 ports, so that no air could pass through the high note ports, and in any case, the admission valve and pipes ducted to the high note ports had long been removed and the apertures on the chest plated over, as here at the Lizard.

ports allow bursts of air into the cylinder to produce a rising note as it escapes through the trumpet. A steady note of 180 Hz is maintained when the speed of rotation is regulated by the governors at 700 rpm.[33]

When the Visiting Committee of the Elder Brethren arrived at the Lizard aboard the *Irene*[34] for the annual inspection in 1900, they '... found a considerable amount of fog at the station, the fog signal being in operation ...' Mr Matthews explained that the lighthouse suffered 'a high average of fog' but there is no record of any further elaboration on the fog signal. On this visit Mr Matthews was more interested in the proposed improvements to the light and recommended to the Elder Brethren that a single flashing light would be more powerful than the existing two lights.

In 1903, the lantern on the western tower was dismantled and the original electrical fittings removed from the eastern tower. 'A new revolving light of very great power visible at a distance of between forty and fifty miles will replace the two existing lights at the Lizard on October 1st', announced an article in *Engineering*.

The new four panel optic was designed to give one flash every 5 seconds but this was shortly altered to the present characteristic of one flash every 3 seconds.[35] The 700 mm focal length lens is still in daily use and though it no longer '... revolves in its liquid metal bed by clockwork'[36] it seems likely that it will continue to revolve in its mercury bath after automation.

The necessary changes to the electrical equipment in the engine house were less material than those made to the towers at that time. As noted, the Hornsby engines were not installed until other refurbishing work on the fog signal began during 1907. A year or two after the new fog signal was completed, a further aid to navigation was made at the Lizard when a submarine bell was installed about two miles south of the lighthouse.[37] The bell was operated by an electric striker connected to the lighthouse by submarine cable.[38]

Over the last ninety years, the magnificent sound of the twin sirens growling up to speed and then filling the air with rich harmonics and reverberating echoes has become as distinctly a part of the Lizard as the sound of the wind and the birds and the sea.

---

[33] The 16 ports are effectively opened and closed 700 times per minute, i.e. 11,200 separate puffs of air escape through the trumpets; divided by 60, the frequency of the siren is therefore about 186 hertz.

[34] *Irene* was a fairly new ship with twin screws and twin funnels - this ship succeeded the earlier paddle steamer of the same name. This ship was sunk by enemy action in 1915 near the Tongue lightvessel.

[35] The Chance Brothers optic revolves clockwise once in 12 seconds.

[36] Stone, John Harris *op. cit.*

[37] The 1910 List of Lights indicates that the submarine bell was to give 5 strokes every 30 seconds. Submarine bells are covered in Chapter 10.

[38] The submarine bell at South Stack was operated by compressed air like those used on lightvessels, but the Lizard appears to have been electrically operated, despite the evidence that compressed air was used and that a cam wheel for the bell was installed on the siren actuating clock.

Chapter 5

Mull of Kintyre

and the early Scottish sirens

A torrential thunderstorm lashes into the narrow courtyard of the lighthouse at the Mull of Kintyre. Two keepers' cottages face one another with a small tower between them at one end. In front of the lighthouse a steep slope runs down into the cliffs and then the sea 240 feet below.

Near the tower, in a corner of the courtyard, a milestone indicates the distance to Campbeltown up over the mountain: '16 miles 1,030 yards'.

The construction materials were all conveyed over the mountain to this isolated site as it was impossible to make a landing by sea on the west coast. A landing place was established on the south-east coast about six miles away and a road to the lighthouse was specially constructed, but each journey was a day's work. From the east side of the mountain a gradual climb to the 1300-foot high moors is followed by a steep slope which drops quickly towards the cliffs on the west. A narrow road cautiously zigzags three quarters of a mile down a 1000-foot descent to the lighthouse. From the top, the coastline and mountains of Northern Ireland can be seen to the south-west: two or three towering anvil-shaped clouds are dragging showers towards us; but in the blue skies to the south-east, the view opens out to the sea and the island of Sanda. To the left of Sanda, in the distance, is the distinctive outline of Ailsa Craig, the top of which is lost in cloud.

The 1786 Act of Parliament which established the lighthouse authorities for Scotland and Ireland authorized the Commissioners of Northern Lighthouses to build four new lighthouses.[1] The first of these was at Kinnaird Head, first lit in 1787. The second lighthouse, at the Mull of Kintyre, was two years in construction and was first lit in November 1788. Built by Thomas Smith, only the thick walls of the tower remain from the original construction. The original dwellings were pulled

---

[1] Kinnaird Head, Mull of Kintyre, North Ronaldsay, Eilean Glas.

down in the 1850s when the cottage on the north side was added. The accommodation on the south side was built in 1883. One or two other buildings have been and gone over the last two centuries and there have been several changes to the lantern.

The tower is a mere 40 feet tall, but the site chosen by Smith is 290 feet above sea level. The white buildings have the distinctive black lantern and cupola of the Northern Lighthouse Board, and window frames, railings and other details are picked out in ochre.

With the rain passed over and the sun now out, it seems the only sounds come quietly from sea level. The peace and quiet outside the lighthouse is transformed into a reverent hush inside, except for the slowly ticking clock in the pale blue entrance room. There are 15 steps up to the lantern and the staircase winds around the weights which are suspended in the open stairwell of the tower. The weights are still connected to the gleaming brass of the clockwork beneath the optic. Although the lighthouse was electrified in 1976 and the clockwork is no longer used, it could quite easily be brought back into service in an emergency.

The original fixed light at the Mull of Kintyre gave way to a flashing character and the most recent optic was designed and installed by David A. Stevenson in 1906. It is a Chance Brothers' optic but Doves of Edinburgh completed the engineering. This information has nearly evaporated from the surface of the brass plaque after 90 years of polishing! Other polished brass gleams with warm tones against the sky-blue wood panelling in the service room.

The engine house, which was built slightly to the north of the lighthouse in 1883, has been converted into a bothy for visiting engineers, but on the cliff top in front of the lighthouse stands the original fog signal horn house from 1883. The hexagonal concrete structure has a maximum internal width of 12 feet and stands 16 feet tall. From the elongated domed roof a large iron trumpet protrudes a further 8 feet and is surrounded by iron railings which have leached rusty stains over the entire building.

Nearly a century after the light first shone at the Mull of Kintyre, the lighthouse was provided with a siren fog signal.

David and Thomas Stevenson, the sons of Robert Stevenson were engineers to the Northern Lights for 20 years from 1855 and were responsible for installing the Daboll reed at Little Cumbrae in 1865. In 1872, the Secretary to the Commissioners of Northern Lighthouses wrote to the Board of Trade and to Trinity House regarding the installation of fog signals at two major sites in Scotland which David and Thomas Stevenson proposed to construct. The two sites proposed were at St Abb's Head lighthouse on the south east coast and at Sanda, in the Clyde approaches off the south east of the Mull of Kintyre.

There was no action taken by the Board of Trade until August of 1873 when the proposal was agreed upon after consultation with Trinity House. It was the view of the Elder Brethren that the results of the South Foreland trials might have a bearing on the Scottish proposals, although they believed that the trials were sufficiently advanced, after three months, to feel confident in recommending Professor Holmes' reed horn, which they felt should be '... permanently classed as a coast fog signal, and adopted for service without waiting for further developments'.[2]

Trinity House gave the statutory consent for a fog signal at St Abb's, but regarding Sanda, the Elder Brethren believed that the proposal should form part of a larger plan for navigation in the 'north channel'. This would require further discussion between the Commissioners of the Northern Lighthouses, Commissioners of Irish Lighthouses and Trinity House.

In the meantime, the Stevensons had seen the Brown siren and corresponded with the New York manufacturer. In 1874, shortly after the conclusion of the South Foreland trials, the Stevensons asked Browns to supply quotations for siren equipment, engines and compressors. It would cost $4900 to supply two 24-inch Ericsson engines, pumps and second-class siren. If 32-inch cylinder engines were required, this would increase the cost to $6900. Alternatively, Browns could provide a steam siren with 25-horse-power steam engines and all fittings at a cost of $3600.[3]

By the end of October 1874 the location of the engine houses had been planned by the Stevensons and the proposed buildings approved by the Board of Trade. The buildings were to be substantially built of fireproof materials. At Sanda the building was constructed in the irregular space remaining on Ship Rock, between the lighthouse tower and the stair tower. The longest walls were 40 feet by 25 feet. The fog signal house at St Abb's measured 30 feet by 25 feet and was constructed to the north-west of the tower and north-east of the keepers' accommodation. The cost of both sirens, complete with hot air engines and delivery and construction of the buildings, exceeded £4000. St Abb's became the first siren in Scotland in 1876 and Sanda was completed later in the same year. Unlike the South Foreland siren, both of the new Brown sirens were sounded by compressed air, and both sounded at a lower pitch. Where a horizontal resonator was used at South Foreland, the new sirens had vertical trumpets similar to the trumpet design of Holmes' reed.[4]

The development of the siren by George Slight and Frederick Holmes has already been considered, and in 1875 the cost of a Holmes' patented automatic siren complete with 32-inch caloric engines, pumps and receivers was £1660.

Lightvessels around British coasts acquired the earliest Holmes' sirens and the lightvessels at Abertay and Bahama Bank, north of the Isle of Man, had Holmes'

---

[2] Letter dated 13th August 1873 from Trinity House to the Commissioners of Northern Lighthouses, contained in 'Correspondence relating to Fog Signals', Parliamentary Proceedings.

[3] A. & F. Brown went on to produce a range of sirens of different sizes, operated by steam and air. In later years their sirens mostly used the automatic self- driven principle, rather than being motor-driven like the early instruments.

[4] A photograph of St Abb's Head from about 1880 shows a small trumpet protruding from the roof of the engine house.

reeds or sirens shortly after St Abb's and Sanda had their Brown sirens. In 1878 David and Thomas Stevenson consulted Holmes about the construction of a fog signal at the Fair Isle.

In 1871 Holmes had installed a reed at Souter Point lighthouse, which had a long air pipe to connect the engine room with the fog signal. In Canada, Holmes had installed a signal that was worked at a distance of half a mile. In both cases the compressed air was conveyed to the fog signal by buried pipework.

There were no further developments regarding the proposed signal at Fair Isle at that stage, but in 1881 the Stevensons again contacted Holmes, this time in connection with plans for Ailsa Craig.

John Wigham[5] submitted proposals to Trinity House to use gas as an illuminant in lighthouses. Town gas was already in use at several harbour lighthouses and the Stevensons, who had taken an early interest in the four stroke oil and gas engines, used gas engines fuelled by town gas near Gourock on the Clyde. Gas engines could be quickly started to begin compressing air for the fog signal, whereas the oil and hot air engines could take up to twenty minutes.

In June 1880, the Board of Trade approved the Stevensons' proposal to use gas for the new lighthouse at Langness on the Isle of Man. Successful trials of gas engines had been undertaken by the Northern Lights in 1879 and the Langness plan included the construction of a gas works to supply the lighthouse. James Keith of Arbroath was appointed to construct his 'patent mineral oil gas works' which distilled the gas from paraffin oil. Two 3½ hp gas engines were installed at Langness, which consumed 20 cubic feet of gas per horse power per hour. The gas holders could store 4000 cubic feet.

Based on the successful operation of the plant at Langness, the Stevensons produced plans for a lighthouse and fog signal station at Ailsa Craig. Gas would be used for the fog signal engines and as the lighthouse illuminant.

The small rocky island of Ailsa Craig rises steeply to 1200 feet and it was not a straightforward site for building a lighthouse and fog signal. The only feasible site was on the south-east shore but this would result in a very restricted range for the fog signal.

Charles Ingrey was the superintendent engineer to Holmes' Caloric Engine & Siren Fog Signal Company. Ingrey and the Stevensons agreed on a proposal for the fog signal, which was accepted by the Northern Lights, but the Board of Trade regarded the scheme with scepticism and were reluctant to approve of the expenditure.

The proposal was to build two fog signals at opposite ends of the island which would be automatically sounded by compressed air. This was to be achieved by

---

[5] John Wigham was responsible for many engineering developments for the Commissioners of Irish Lights. See Chapter 7, note 3, relating to gas illumination, Tyndall and Douglass. Wigham also developed an air turbine-driven siren.

pumping air over 1¼ miles of pipework from the engine house. Piped air precedents had been far less ambitious and to convince the engineers representing the Board of Trade, Ingrey set about a full scale demonstration of the feasibility of the proposal.[6]

Due north of Ailsa Craig is the small island of Pladda off the south of Arran. Pladda lighthouse had recently been equipped with a Holmes' reed horn, caloric engines and compressors, when, in 1882 or early 1883, Ingrey disconnected the fog signal from the compressor and then reconnected them both via 2760 feet of piping laid around the island! This demonstration was sufficient to convince the Board of Trade.

The work on Pladda, undertaken in about a fortnight with the assistance of the keepers and farm labourers from Arran, was not, however, indicative of the actual difficulties encountered in the ambitious work at Ailsa Craig. The building programme at Ailsa Craig, which included the lighthouse, gas works, accommodation, engine house, fog signals and pipework, took over three years to complete.

During this time the Caloric Engine & Fog Signal Company was additionally invited to install a siren at the Mull of Kintyre lighthouse. In comparison with the work undertaken at Ailsa Craig, construction of the Mull signal was relatively simple and the fog signal was in service in 1884. The new fog signal at the Mull of Kintyre was the first two-note siren in Scotland.

Whatever success Holmes' company appeared to be enjoying, with contracts for fog signal equipment from various lighthouse authorities around the world, the Register of Defunct Companies shows that in April 1885 the company received court orders to wind up the business. In June 1887 the company was dissolved and Holmes, in poor health, retired to the south of France.

The Pulsometer Engineering Company[7] bought Holmes' patents and also took on Charles Ingrey as their senior engineer. The work in hand at Ailsa Craig continued throughout 1884 and 1885 and the lighthouse and fog signals were completed and brought into service in 1886.

The gas works constructed on Ailsa Craig was larger than the one at Langness and could produce up to 10,000 cubic feet of gas from 100 gallons of paraffin oil. There were two gas holders, each with a capacity of 10,574 cubic feet and a full gas holder would keep four engines running for 19 hours. The gas was collected and cleaned in water and elaborate drainage was laid to collect the necessary supplies.

---

[6] Many of the Scottish lighthouses later used air pipes over some distance and the practice was also used in France in the 1880s and 1890s.

[7] The Pulsometer Company had grown from its works in Nine Elms, London, since 1879, where it produced the 'pulsometer' used for pumping bilge and ballast tanks. The pulsometer dates from 1875. In the early 1890s The Pulsometer Engineering Company had works at 35 Robertson Street, Glasgow as well as 63 Queen Victoria Street, London.

Typical equipment supplied by the Caloric Engine and Siren Fog Signal Company about 1883, consisting of two Buckett engines and a two-note siren.

Four keepers were required at Ailsa Craig because of the additional work involved in producing the gas. It was the responsibility of the Principal Keeper to supervise gas production, besides the operation and general maintenance of the lighthouse and fog signal equipment. The Principal Keeper also had to keep watch during fog, unless the period of fog exceeded eight hours when he was permitted some sleep!

Gas production entailed preparing the coal furnaces and heating the iron retorts for up to five hours until they glowed bright red. The retort doors and cleaning ports all had to be sealed with a mortar of fire clay and the gas holders had to be filled with water, '... breaking the ice two or three times a day' in frosty weather![8]

Oil was to be admitted at a rate of about four ounces every minute and the entire gas distilling process demanded vigilance for as long as necessary to refill the gas holder, which could be up to five hours. If the temperature of the furnace was kept too high or too low the process suffered. The inflow of oil had to be measured every half hour and the water levels checked in the washers and gas holder. There could be problems of gas leakage or gas pipes blocking and if the fog signal engines happened to be running at the time, the process was more complicated. Once the gas holders had been filled, it was important to begin cleaning the retort soon after

[8]   *Station Instruction Manual for Keepers* (Ailsa Craig).

71

the oil was shut off and the fires knocked out, to prevent problems caused by the build up of tar and sooty deposits.

The gas was used as the illuminant at Ailsa Craig as well as for the fog signal engines. There were five 8 hp engines made by Crossleys of Manchester. The compressed air demands of the system required about 30 hp and four of the five engines were run in rotation. With four engines working the compressors at 75 psi, about 540 cubic feet of gas were used per hour. The gas was mixed two to one with air for a better burn and ignition was provided by a pilot light, not an electric spark. The air compressors ran at 60 rpm from overhead belts off the engines. There were two double cylinder compressors with 10-inch cylinders and 18-inch stroke capable of delivering 218 cubic feet of air per minute at 75 psi.

Inside the engine house were two air receivers open to the compressors and these charged the receivers at the horn houses at each end of the island prior to the signals being sounded. From the engine house, the north horn house was 1130 yards away and the distance to the siren at the south of the island was 830 yards. Two and half inch diameter wrought iron pipework was buried in a trench two feet deep to link up the system. Laying the pipes had been difficult, especially in poor weather and involved blasting the granite in many places. Once the system was assembled, complete with wrought iron bridges over some parts of the route, high pressure air was pumped through the pipes to check for leakage at the joints before filling in the trenches with concrete to protect against rock falls and the elements. Where there were dips in the course of the pipes, Charles Ingrey had installed syphon boxes to collect water which might otherwise freeze and damage the pipes: compressed air contains water vapour. It later became the keeper's job to periodically open these valves, with the system under pressure, to blow out any water which had collected. Once the pipework system was completed, the concrete-filled trench served as a pathway to each horn house.

Both horn houses were large, conical, concrete structures built above the high water mark at the foot of the precipitous cliffs. Inside each, the pipework fed into a nine foot high iron air receiver with a capacity of 150 cubic feet. Installed on top of the receiver was the admission valve and the siren and a twelve foot brass trumpet, which protruded through the middle of the domed roof.

On earlier Holmes-designed reeds and sirens, the signal was sounded by the simple use of a weighted lever which was tripped over when air pressure in the receiver exceeded the resistance of the weights and opened the valve. Other systems used cam wheels, driven by the engine, to open small actuating valves which, in turn, opened the main air admission valve to the siren. With the introduction of the distinctive two-note characteristic fog signals, the latter was the more useful system, but at Ailsa Craig the system had to control two fog signals at a distance.

Ingrey devised a system which used clockwork to actuate the fog signals and an integrated self-winding mechanism which was operated by the compressed air. A weighted lever was still used on the receiver in the horn

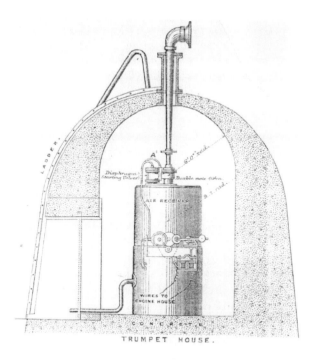

LADDER.

Diaphragm
Starting Valves

6'.0" rad.

Double note Siren.

8.3 rad.

AIR RECEIVER

WIRES TO
ENGINE HOUSE

CONCRETE

TRUMPET HOUSE.

The south horn house at Ailsa Craig. Courtesy Institution of Civil Engineers.

house, but the operation was more complicated. At the opposite end to the weight, the lever had a toothed quadrant which engaged a small pinion wheel. When the lever was raised the quadrant turned the pinion which turned a larger chain wheel on the same spindle. The weight lever itself was raised by a small piston operated by the rising air pressure in the receiver. A second chain wheel was linked to the first by an endless chain, from which was suspended a weight. At the extent of the movement of the pivoted weight lever, the downward movement of the quadrant also tripped a connecting rod to release the second chain wheel, which began to rotate by means of the falling weight on the chain. The second chain wheel was on the same spindle as two cam wheels, the cams of which pressed against actuating valves: as the wheels turned, the valves released air through the admission valve and sounded the siren.

The finer points of the self-winding mechanism involved using ratchets to prevent the chain wheels from running down and a train of other geared wheels attached to a vane which regulated the running of the chain wheel and cam. The vane, as used in clock striking mechanisms, could be adjusted by the keeper when the siren was timed for accuracy.

> Each time the pressure in the receiver approached 70 psi the piston moved the weight lever and the clockwork moved the cam wheel through a quarter turn, to sound the siren. The reduced pressure in the receiver allowed the piston to release the weight lever and the winding machinery stopped, ready to repeat the cycle in three minutes' time.

The two fog signals at the north and south of the island were designed to sound alternately at three-minute intervals, or with a 90-second interval between them. This regular operation could have been vulnerable to fluctuations in the air pressure in the system, and a further safeguarding feature was devised which prevented the automatic operation of the compressed air and clockwork from sounding the signal prematurely.

An electric circuit was briefly completed every 90 seconds alternately between the engine house and the north or south horn house and an electromagnet triggered the piston on the weighted lever. The signal was normally sounded at about 70 psi, but the weight lever was designed to trip at a pressure of 40 psi. Therefore, when the electric circuit was made and the electromagnet opened the air to the weight lever piston, the signal sounded almost immediately. The electric circuit[9] was switched on by a cam wheel driven directly by the engines.

Such elaborate actuating machinery was unnecessary at the Mull of Kintyre due to the relative proximity of the engine house and horn house. Here the actuating clock was simply a cam wheel in the engine house, mechanically driven by the hot air engines. The cam wheel rotated once every four minutes and the cams opened the actuating valve which communicated with the admission valve in the horn house through a quarter inch pipe. The main air pipe and the actuating pipe conveyed air between the engine house and the horn house, a distance of about 140 feet.

The admission valve, or 'starting valve', was six inches in diameter and was effectively closed by the pressure of air in the system. The siren was designed to operate at about 75 psi, as were those at Ailsa Craig. It was reckoned that the pressure exerted on the admission valve open to the receiver was about a ton.

The 1883 Mull of Kintyre siren was a Holmes two-note instrument and appears to have had a brass trumpet similar to those at Ailsa Craig. Inside the horn house, an iron receiver over ten foot tall stood in the centre of the building with the siren and valves mounted above. Air pressure kept the valves closed until tripped by the opposing air pressure from the actuating valves which regulated the duration of the blasts.

The two-note instrument at the Mull was a single automatic siren and the characteristic was two seconds for the high note, silent for two seconds, two seconds for the low note, 234 seconds silent.

---

[9] The battery-powered circuit was connected to the horn houses by copper wire coated in bitumen and encased in wooden troughs which followed the air pipes. The circuit completed a secondary circuit at the horn house which opened the air valve to the piston.

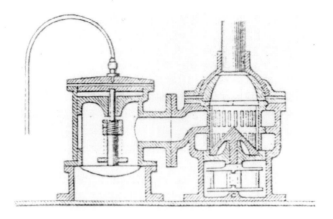

Section through two-note siren and high note admission valve. The valve opens when control air from the timer presses the valve stem down. Courtesy Institution of Civil Engineers.

One admission valve sounded the high note through a duct to the 32 ports at the top of the siren and the other valve sounded the low note, which was produced by 14 larger ports in the lower part of the cylinder. The low note sounded at 280 Hz and the high note at 640 Hz. The Holmes siren rotated at 1200 rpm and the centrifugal weights which spun out to retard the rising notes of the siren were factory fitted, whereas the inventor of the siren governors, George Slight, used an adjustable screw which enabled the sirens to be tuned at the lighthouse. According to company publicity the Holmes sirens were deliberately tuned to be discordant to help distinguish them from other sounds!

The characteristic of the north siren at Ailsa Craig was one five second blast every three minutes and the south signal was a high note for two seconds, silent for two seconds, low note for two seconds, silent for two seconds, high note for two seconds, silent for 170 seconds. The 90-second interval between the north and south signals additionally helped to prevent mariners from confusing them.

The Ailsa Craig trumpets were tilted slightly downwards and could also be turned to face into prevailing winds. The copper horns had hinged covers over the trumpet mouths, which were subsequently used at other Scottish sites, designed to keep out rain and spray. The covers, which blew open with the first siren blast, were supposed to be replaced by one of the keepers as soon as possible after the fog had cleared.

Much of Ingrey's pipework was subsequently replaced, as were the sirens, and new air receivers were later installed beside the horn houses. The gas engines were replaced by diesels and both sirens were discontinued in 1966 when a supertyfon air signal was installed near the lighthouse.

The engine house at the Mull of Kintyre was originally fitted with two Buckett caloric engines, typical of those Charles Ingrey would have installed at that time. He probably felt that these engines would have been suitable for use at Ailsa Craig too! The engines had upright double cylinders and produced eight brake horse power at 60 rpm. Engineering drawings signed by D.A. Stevenson[10] and dated 20th October 1897, indicate that three oil engines replaced the two original Buckett hot air engines. The single air receiver from 1883 was to be supplemented by a further three receivers inside and two outside the engine house in 1897. At the horn house itself, another air receiver 4 feet 6 inches in diameter, was added outside the building and the original four-foot diameter receiver inside was moved to accommodate the new siren standard. New water tanks for cooling the engines were also built at this time. The refurbishment was completed in 1898 when the Admiralty List noted a change to the characteristic. The new fog signal at the Mull of Kintyre gave two five-second blasts, first high and then low, every 90 seconds.[11]

The engineering work was undertaken by Steven & Struthers of Glasgow, to the specification of David Stevenson. The work did not require substantial alterations to the 1883 horn house, although railings were added to the roof to enclose the new cast iron trumpet. The trumpet was 3 foot 6 in across the mouth and had a total length of 22 feet, much of which extended vertically down inside the horn house to the standard, which contained the admission valves and siren chest. The standard was 3 foot high and 4 foot 9 in long, and was installed directly below the horn once the original air receiver had been moved from the central position. The standard remains on site today along with the horn, although the admission valves and siren have long gone.

The 1883 Holmes trumpet could be turned manually to face into the wind, but the new Stevenson trumpet had a worm wheel and crank handle near the throat to help move the heavy horn. The greater quantity of air used by the new siren is indicated by the increased number of receivers and the new six inch diameter air main which linked the engine house to the horn house. The new siren sounded more frequently than the original one and its ports were probably larger. The diameter of the siren chest was 10 inches, and the overall diameter of the siren might have been about 7 inches. The admission valves were 13 inches in diameter.

Hearing the siren off the south of the Mull was sometimes impossible and this finally led to the construction of a new horn house (ordered by David Stevenson) on the clifftops nearly a mile to the south of the lighthouse, in 1913. As at Ailsa Craig, air pipes were laid to connect the siren to the compressors in the engine house, and traces of the four-inch-diameter iron pipework can still be seen emerging from the ground in places. The 10-foot lengths of pipe follow the undulating contours and are buried in places and elevated in others.

[10]   Engineering drawings held by the NMRS, Royal Commission for the Ancient and Historical Monuments of Scotland.

[11]   An official enquiry into the wrecking of the paddle steamer *Signal* in 1895 raised the question of the long interval between the blasts of the earlier siren at the Mull of Kintyre.

The original Mull of Kintyre horn house after the 1898 modification, as it appears today.

The new siren was operated at 25 psi and sounded a low note of 130 Hz.

To accommodate the air quantity used and to equalize pressure as the signal sounded, six receivers were installed at the horn house and another four were added at the engine house. The design of this siren was typical of those which protected the Scottish coasts and with only a few modifications, including new engines in 1941, the siren remained in use until 1976, when electricity arrived at the Mull of Kintyre and the engines were removed.

A small room at the base of the tower, formerly an oil store and then a wash house, became the new engine room and contained the high-powered electric Hydrovane compressors which pumped air along the cliffs to the horn house. Small bore flexible pipework replaced the iron pipes and conveyed the air at much higher pressure. In the horn house were three Air Chime diaphragm horns which blasted the Morse letter N every 90 seconds.

When I was at the Mull in 1994, work was already underway to lay an electric cable to the horn house and in due course the Air Chimes were replaced by an electric fog signal. Other preparations towards the automation of the lighthouse culminated in the summer of 1996 when lighthouse keepers finally left the Mull of Kintyre, after a continual presence of two centuries.[12]

---

[12] With other automatic Scottish lighthouses, the Mull of Kintyre lighthouse is monitored from NLB Headquarters, George Street, Edinburgh.

Chapter 6

Beachy Head
and the explosive fog signal

The new lighthouse at Beachy Head was begun in 1900 and constructed at the foot of the cliffs on a hard chalk foundation. Fairly deep water runs quite close to the base of the cliffs, though the bedrock on which the lighthouse stands runs out for about half a mile forming a slight ridge. The tower stands about six hundred feet from the base of the cliffs, virtually on the edge of low water spring tides.

Thomas Matthews was responsible for the construction of the lighthouse, which replaced James Walker's 1828 light at Belle Tout, a mile or so west of the new lighthouse. The highest cliffs at Beachy Head are over 500 feet and cliff erosion results in periodic rock falls. Belle Tout, on the clifftops, was vulnerable to erosion, but even though this was reputedly a consideration when deciding to build a new tower at the foot of the cliffs, the main reason for deserting the old lighthouse was that, as at Lundy and the Needles and elsewhere, the low cloud forming on the clifftops often obscured the light.

The original clifftop Needles light was demolished in 1859, on completion of the tower at the end of the Needles rocks, but the old tower at Lundy remains, as does Belle Tout, having, thus far, avoided collapsing over the cliffs.[1]

Matthews' task was less difficult than that of the Douglasses and the Stevensons, although he used the well-established construction technique of interlocking granite blocks, used at the other rock towers. This was one of the last tower lighthouses built in this way. A coffer dam enabled excavation below the water line for the foundations and an iron platform was installed adjacent to the site.

A steam winch linked the platform with the clifftops and the aerial ropeway transported the granite blocks down to the site. Some 3660 tons of Cornish granite was lowered to the building works on 6-inch cables. The construction platform had a 5-ton steam crane to assist the positioning of the blocks in their courses, a total of

---

[1] In 1999 the whole tower was raised on rollers and moved back from the unstable cliff edge.

Beachy Head lighthouse.

75 including the foundation blocks. The blocks were prepared on the clifftop but were cut at the De Lank Quarry near Bodmin and transported to Eastbourne by rail, before finally being hauled up the clifftop by steam traction engines. The last stone was set in February 1902.

In the following June, Matthews and Lord Rayleigh, who succeeded Tyndall as the Scientific Advisor to Trinity House, visited the Chance Brothers' factory to inspect the new optic for Beachy Head. The optic was reassembled in the lantern and first exhibited on October 2nd, 1902.

To distinguish the lighthouse as a daymark visible against the white cliffs, the tower had a black band and black lantern, the remainder of the tower being left as natural grey stone. This must have not been entirely satisfactory since the more familiar red band and red lantern has conspicuously stood out since World War Two. The previously bare stone is now painted white.

Supply and relief boats delivered oil, water, explosives and sundry other items to the landing stage at the base of the tower, but it was possible to walk to the lighthouse along the foot of the cliffs at low water spring tides.

From the landing stage, gun metal dog steps climb up the tower wall to the main entrance about fifteen feet up. There are huge gun metal doors as at the other rock

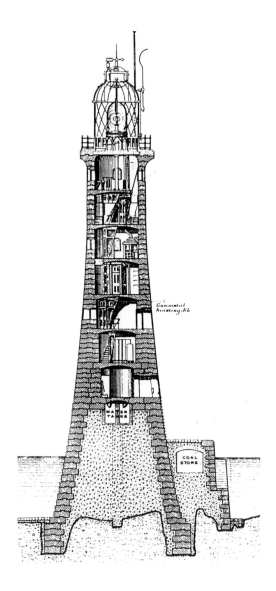

Section drawing of tower showing the explosive jib, lantern, service room, bedroom, kitchen, store room, crane room, oil room and entrance room. Courtesy Institution of Civil Engineers.

lighthouses, but such doors are secured from the inside, so here, additional doors have been installed since automation. The door opening is nearly 40 feet above low water spring tide (LWST) and the entrance room is nearly nine feet into the granite; more a short corridor than a doorway.

To reduce problems of condensation and damp, building maintenance is assisted by heating the inside of automated lighthouses and it is rare to find a cold lighthouse. After the walk it feels positively hot inside the entrance, but in the past the lower rooms of tower lighthouses were usually cold and damp.[2]

The entrance room is the smallest in the lighthouse and on the floor the exquisite pattern of the interlocking granite with its lead pointing can be appreciated as also on the slight vaulting of the ceiling. The granite is finely cut but coarse enough in texture for the mica to sparkle. The floor has two gun metal manhole covers over the freshwater tanks sunk below the entrance room. The manhole cover on the landing outside seals the old coal store. This lighthouse had a Cornish range until 1962 when a Rayburn was fitted. The water tanks can be checked with a measure and a hand pump supplies water to the header tank in the kitchen. This lighthouse also has a toilet which uses collected rainwater. The remaining fittings on this floor include a salt water pump, ropes and lifesaving equipment, flags, flares and a lifebelt with 'Beachy Head Lighthouse' clearly painted around its rim.

On the left, the painted cast-iron steps with brass handrail lead steeply up to the oil room. There are only 15 steps between each floor. A curved door opens off the steps to the oil room and inside are eight segmented floor to ceiling tanks filling the space, 5 × 180 gallon tanks and 3 × 140 gallon tanks. It smells like the bilges of a boat in here and it is obviously not heated, so it feels cold and the diesel stinks! The fuel levels are checked with a long dipstick, though it is hard to manoeuvre the measure into the tank with the limited head room. As the emergency generator is not often called into action, except for the regular checks, the fuel level doesn't fall much between visits.

This room originally had 3 × 140 gallon tanks for storing the paraffin used by the incandescent oil burner in the lantern, and there was also a handpump to keep the pressure up to 80 psi in the pressure tanks. The two pressure tanks, one of which was a reserve against mishap, supplied oil to the vaporizer up in the lantern.

The internal diameter of the rooms increases to a maximum of 15 feet further up the tower. The next room is the crane room. This room has a door opening situated above the main entrance overlooking the landing stage. Securing this door are two massive bronze bolts, over two inches in diameter, and at least 18 inches long. With these doors open, supplies could be lifted aboard with the gun metal hoisting jib which protrudes out over the door.

Some of the storage cupboards in the crane room have been reappropriated and a shower unit has also been installed on this floor. The battery chargers quietly buzz. Beside the grey metal chargers, which stand immediately on the right at the top of the stairs, there is – on the left – a work bench with a vice and toolbox. There

---

[2] With the introduction of solar panels at rock lighthouses, it may be that the lower rooms of the towers will again become cold and damp as the solar panels will not produce sufficient energy for heating purposes.

are a dozen batteries in the small storage space behind the chargers. Walking past the door and climbing the steps over the shower, the next room is the main storeroom.

Like many isolated towers, Beachy Head had an explosive fog signal. Explosive charges were stored in the magazine, a wooden-lined steel cabinet in the storeroom. Detonators were probably stored in the room below. There is now a freezer in the storeroom but before its arrival, vegetables were stored on the window ledge between the window and the shutter to keep them cool. The window space now has the exhaust pipe from the standby generator venting through it. The cupboards still contain all sorts of interesting items and the room has a tool kit and the 18 bhp Lister generator. The lighthouse has been on mains electricity since 1975, but in the event of a power failure the generator automatically cuts in.

The kitchen or living room is on the next floor. This room is fifteen feet in diameter and has two windows at right angles to the door. The windows look up and down the coast, not quite east and west. Many original fittings in the kitchen have given way to periodic improvements over the years and there is a little extra space where the Rayburn, and its predecessor from 1902, the Cornish range, once stood. The solid fuel stoves were used for all cooking, heating and hot water and the chimney also warmed the bedroom as it passed up the tower. The kitchen now has an electric cooker, microwave, 'fridge and electric heating. There is a table and chairs and there are original curved stowaway cupboards which double as seating. Above the seats is a bookcase shaped to the curvature of the walls. Tucked in the space above the stairs down to the storeroom is a fitted cupboard complete with a desk and drawers. Above the west window is the freshwater header tank supplied from the tanks below the entrance room.

All of the window frames above the crane room are made of oak, but the lower rooms have gun metal frames. The glazing is half-inch thick plate glass. The lower windows also have gun metal storm shutters. Peering out of the kitchen window, the thickness of the walls is judged to be two feet.

The sound of the electric optic drive is audible from the bedroom above the kitchen, but it is not loud or unpleasant. This constantly present sound was not a familiar one to the keepers, but only to those periodic visitors who now clean and maintain the lighthouse: since automation in 1983, the optic has rotated day and night.

The bedroom door curves with the radius of the inner wall and inside the bedroom are six curved bunks with fitted floor-to-ceiling storage space. There is just enough room for a small washbasin beside one of the windows. When manned, the three top bunks remained available for visiting mechanics unless preferred by one of the keepers. Falling out of bed could have been painful; a nine foot drop to the granite floor! The severity of the granite walls is mellowed by oak lining inside the bunks, which can be curtained off for privacy. The drawers and cupboards below the bunks form interesting shapes. The bunks do, indeed, have curved mattresses and I believe those keepers who say that making the bed was awkward! However, I have a sleeping bag and stretching diagonally across the curve causes me no problem. The other five bunks have no window but mine does and I look forward to watching the view of the cliffs picked out by the lighthouse beams if I can't sleep.

The character of Beachy Head is two group flashes in 20 seconds. The beams sweep from right to left past the window and on over the cliffs, picking out clumps of vegetation and cracks in the chalk. The nature of the optic design and the lantern screen toward the land is apparent from the progress of the two spots of light as they pan from the sea over the cliffs: the first spot fades away at a point on the cliff which is several tens of feet before the second. At first, the stars can be seen above the clifftops and there is a faint orange glow from Eastbourne. The beams light up the cliffs and reflect in the water, which itself looks rather milky. It is difficult to see that the beams are radiating outwards horizontally from 20 feet above my head and instead, they appear as parallel vertical bands wiping across the window.

The sea is only heard very faintly as the tide rises and the sound of the optic drive is almost soothing. A breeze picks up which manages to circulate the air in the tower and it starts to whistle under the curved door of the bedroom.

When he was the Scientific Advisor to Trinity House, Michael Faraday devised the essential ventilation for lighthouses when they all used oil lamps. Without adequate ventilation, the lamps burned poorly and condensation built up in the lantern, further diminishing the range of the light. The oil lamps have long gone but the ventilator design in this lighthouse is obviously still working!

The fog signal is an electric diaphragm horn situated on the lantern gallery nearly 30 feet away from my bunk, but those 30 feet don't do much to diminish the 130 dBA produced at source! It is not easy to sleep through the six-second blast every 30 seconds, but in any case I want to make a sound recording of it, along with that of the Royal Sovereign, which is plainly audible 10 miles away. At this range the diaphone is sweet, but the emitter above is very loud and transmits other peculiar frequencies through the granite. It is 2.00 a.m. and none of us are asleep so I'll go and make a cup of tea and get that recording of distant Royal Sovereign from the kitchen window.

The service room is more cluttered now than in 1902, since many of the original fittings remain and there are, in addition, two large grey metal cabinets containing equipment for the fog signal and telemetry. When the lighthouse was built, Marconi had just effected radio communication between Cornwall and Newfoundland – now telecommunications enable automatic monitoring of lighthouses on computer at the Operations Control Centre in Harwich.

Between the two grey cabinets is a large wooden frame containing several triangular and diamond-shaped replacement glazing panes for the lantern. The curved glass is ⅜ inch thick.

The weight tube is a green-painted cast-iron column in the centre of the room. Inside are the weights, totalling 480 lbs. Clockwork, situated in the lantern room above the weight tube, used to provide the power to rotate the optic. The suspended weights fell about eight feet in an hour and rang a bell fixed to the weight tube to alert the keeper on duty: the clockwork needed winding every hour. A keeper could be dismissed for letting the weights run down to stop the rotation of the optic.

The keepers' duty was to maintain the light in first-class condition and additionally to start the fog signal when necessary. At sunrise the lamp was extinguished and preparations for the evening could begin. Other responsibilities included keeping regular meteorological records and noting any irregularities in neighbouring sea marks – buoys, lights and fog signals. Distress rockets could be fired from some lighthouses to alert coastguards to shipping in difficulty, although keeping a lookout was not, strictly-speaking, part of the job. At Beachy Head, however, there was a direct telephone line to the coastguard station at Eastbourne, which was also a Lloyds signal station.

The lighthouse service regulations laid down specific rules for keepers and expressly forbade sleeping on duty, although it was permissible to have a chair in the service room.

Here at Beachy Head, there is a desk and chair by the east window and a large curved cupboard. Evening light shining through the west window glows on the wood to warm up the otherwise grey and green room. The cupboard is glazed above, whilst opening the doors below reveals a lead sink. Rainwater collected from the lantern roof was used for cleaning purposes and keepers spent hours cleaning and preparing lamps here. The cupboard now contains spare electric lamps and other electrical components plus manuals and spare bottles of mercury. Above the desk is a sea chart of the area and nearby a table of the lighting up times. A brass plaque on the wall above the glazing stand provides information about the inauguration of the lighthouse in 1902.

The final flight of steps, from the service room to the lantern, is open and leads through the iron floor which is supported by huge radial cast-iron girders. Inside the lantern is the impressive Chance optic, which consists of three double panels.

The first order optic completes one anti-clockwise rotation every minute to produce two flashes every 20 seconds.[3] The beams are now produced by a 400w mercury iodide lamp which has a bluish light.[4] The original illuminant was a Matthews-designed incandescent oil burner which used vaporised paraffin under pressure of 54 psi. The sound of the oil burner roaring and the clockwork drive would also have been complemented by the station clock positioned on the screened section of the lantern. On the face of the clock, the lighthouse character and station name were clearly marked. If the clockwork optic drive failed, the optic had to be

---

[3]   Chance Brothers' specifications indicate that the character was one flash of 0.5167 second duration followed by an interval of 2.8166 seconds. The second flash (0.5167 seconds) was followed by an interval of 16.15 seconds. The change to the illuminant now produces a flash of 0.23 of a second with a short interval between flashes of 3.07 seconds and a long interval of 16.47 seconds between groups.

[4]   When the lighthouse is next refurbished the lamp will probably be replaced with an even smaller lamp: some recently automated lights now have a mere 35 W lamp!

Fog detector and emergency light at Beachy Head.

rotated manually by the keeper who kept an eye on the identifying marks on the lens panel and on the station clock, to maintain the character of the light.

The lantern is 14 feet in diameter with helical glazing bars of gun metal supporting the glass and the weight of the roof. Standing 3 feet 10 inches in height from the lantern floor is the pedestal which supports the glazing bars. This is made up of sixteen cast iron segments, each with a ventilation panel and with one providing door access to the gallery. Clambering out on to the lantern gallery through the heavy door affords an impressive view of the coastline. On a bright sunny day the brilliance of the red paint against the white cliffs is dazzling, but the milky turquoise sea produces an even more vibrant contrast with the red.

On the gallery railings is the fog signal and adjacent to it is the fog detector. The fog detector is a Stone Platt FD3, which comprises a xenon tube and a capacitor, which is fired in a similar way to a flash gun. At the top and bottom of the unit is the transmitter and receiving element respectively. The amount of light reflected back into the receiving lens gives a measure of the opacity of the atmosphere and the sensitivity to the 'back scattered' light can be adjusted so that the fog detector switches on the fog signal when the visibility falls below, say, one mile. The fog detector makes four to seven sample readings at six-second intervals every four minutes. Once the fog detector has triggered a signal, it continues flashing and measuring 'back scatter' until the reading is less than the reference level. If this is the case, it makes further readings every four minutes until three consecutive clear groups of flashes are obtained when the fog condition is cancelled.

The fog signal is an ELG 500 pure tone electric emitter made by Wallace & Tiernan and it was installed in 1976. Orientated south-east on the gallery with baffles on the railings, it has four directional emitters and a nominal range of four nautical miles. There are two units of two horns. The fog signal drive unit, in the service room, operates the coder and regulates the flow of current to the emitter stack on the gallery. The two emitters have two steel diaphragms which are electrically vibrated to oscillate back and forth at twice the frequency of the current. Both units are synchronized in a master and slave arrangement to prevent partial cancellation, 'beats' or other diminishing interference between them. The frequency of the signal is approximately 500 Hz.

The rear of the lantern is screened and there are remains of brackets for the explosive jib. The explosive fog signal at Beachy Head was one report every 10 minutes when the station came into service, but in 1903 the Order Book indicates the change to one every five minutes. A few pages further on, the Order Book notes that the lantern roof and ventilator were heavily pitted by the explosions which occurred just a few feet above! Beachy Head was one of the last rock lighthouses to retain the explosive fog signal, which remained in service until 1976. Most other rock stations had supertyfon air signals installed in the late 1960s and early 1970s.

The explosive fog signal was highly effective and took up very little space on a crowded rock lighthouse, but explosives were also used at some shore lighthouses too. In essence, an explosive charge was fitted with a detonator and suspended from a pole above the lantern. It was manually fired by a small electric current, according to a clock.

If fog set in, the keeper on duty had to run down the tower to collect some charges and detonators and en route back to the lantern, make sure that the stove and the chimney were well closed down – it occasionally happened that the explosions shook soot into the kitchen and all over the place! Multi-wick burners in the lantern were quite resilient to shocks but with the new incandescent oil burner at Beachy Head in 1902 it was first thought that the bangs might damage the delicate structure of the collodion mantle. Despite the shocks to the lantern roof, however, it did not affect the illuminating apparatus.

In the lantern, the keeper now had an arm full of charges plus the detonators, plunger and timer. The four ounce tonite charges were about five inches long and originally made by the Gun Cotton Company, but latterly ICI produced the 'Nobel Sound Signal'. The separate detonator was pressed into the charge and the trailing wires tied around the middle. The primed charge could then be suspended from the jib by the copper spring clips which made the electrical contact. The jib was designed by James Douglass, who described its operation as follows:

> ' ... the jib is ... raised and lowered by a worm wheel and pinion, worked by a hand wheel inside the lantern. When the jib is depressed, the lower end reaches near the gallery outside the lantern, where the light keeper suspends the charge or charges of gun cotton with their detonators already attached to electric cable or cables, which

A keeper attaching a charge to the jib at the Bishop Rock lighthouse in the 1930s. Courtesy of Gibson of Scilly.

> *are carried from the end of the jib to a small dynamo-electric firing machine, placed in the lantern. After suspending the charge or charges, the light keeper returns to the lantern, where he raises the jib to the upper position where all charges are fired nearly vertically over the glazing of the lantern, and thus without causing any fracture of the glass'.*[5]

The tubular steel jib with its cow horns[6] is familiar to ex-keepers and this design was used at Bishop Rock from the 1890s. For some reason, an earlier design was used at Beachy Head, which had a wrought iron framework with the ends broadening out to a separation of about eight feet. However, this jib was replaced by the tubular steel type within ten years.

[5]   Trinity House, *Miscellaneous Correspondence and Reports*. Unpublished papers and documents.

[6]   Douglass describes the improved jib installed at Bishop Rock in the *Minutes of the Proceedings of the Institution of Civil Engineers*, Vol. 108, 1891–92. 'The apparatus consists of a hollow steel crane attached to the lantern and actuated by a worm wheel and pinion. For the attachment of the gun cotton charges, the crane is lowered, and two charges are suspended to the firing cables which are connected with a dynamo electric exploder placed within the lantern. After suspending the charges, the crane is raised to its upper position, and contact made at five minute intervals. The charge consists of four ounces of gun cotton into which a fulminate of mercury detonator is placed before suspension ... A gun cotton explosive fog signal ... is now in successful operation at five Trinity House rock lighthouse stations ...'

Most keepers described the job of priming, loading and firing the explosive signal as unpleasant and, at best, interminably boring. Having attached the charges,[7] one on each side of the jib arm, the keeper wound the jib into position with a hand crank and this process made electrical contacts with the plunger. On hearing the bell of the timer 'ding', the first charge would be fired. After the appropriate interval the timer would 'ding' again to prompt the keeper to fire the second charge, after which the jib was wound down, the remaining wires discarded from the clips and two more charges attached. The second charge was originally a precaution against the first not firing. The interval between explosions was typically five minutes, as at Beachy Head, but it was three minutes at some lighthouses and at others the character required both charges to be fired off with only a brief interval between them, every five minutes. When the jib was wound down it broke the electrical contacts as a safety measure, but nevertheless there were some accidents with this kind of signal. At one lighthouse the contacts were made when the jib was not in the correct position so that several glass panes required replacing! There was a similar but more serious accident at an Irish lighthouse, where a keeper lost an arm in the premature explosion.

With the jib down, the keeper took two prepared charges out on to the gallery in the fog or rain or snow, attached the charges and returned to the comparative warmth of the lantern. At some lighthouses there was a special hatch[8] beside the point at which the jib came down, so that the keeper only needed to reach through to attach the charge and not venture outside each time, unless something went wrong.[9]

At one lighthouse, the prevailing wind usually succeeded in blowing out the lamp every time the gallery door was opened, so the last thing on the keeper's mind was firing the charges off to the dictate of the clock!

With several charges prepared, but no more than were likely to be required for safety reasons, the keeper probably had enough time to go down to the store for some more charges and drop into the kitchen to check the teapot.

On the coasts of the British Isles, about eighteen lighthouses used explosive charges for a fog signal, including some harbour lighthouses. The Northern Lighthouse Board only used explosive charges at Skerryvore and Chicken Rock, although Moyes gun and similar gas explosive signals were used elsewhere by the NLB and the Irish Lights.

The German scientist Christian Schonbein, following experiments in the 1830s, produced gun cotton in 1845 by treating cotton with nitric and sulphuric acids.

---

[7] Two insulated copper wires ran to the end of the jib, ending in two brass spring clips into which the ends of the wires protruding from the detonator were inserted.

[8] The keepers at Beachy Head were quite fortunate in this respect – there was a hatch which enabled the keepers to attach the charges without venturing out on to the gallery. At Skerryvore, there was a little iron shelter built on to the gallery which extended from the lantern: the keepers slid open a little port to attach the explosives to the jib.

[9] If the whole thing should jam, and it happened more than once, the unfortunate keeper had to clamber on to the lantern roof to try and free the jib and liberally coat the pinion with grease!

Frederick Abel, who had followed Michael Faraday as lecturer in chemistry at the Royal Military Academy at Woolwich, developed ways of stabilising the product in 1868 and later suggested that Tyndall might try gun cotton in the 1874 gun trials. The trials confirmed the recognised view that gun cotton gave a sharp report which compared favourably with the guns used. The guns often produced a prolonged murmur rather than a distinct report.

Officers from the Royal Artillery and the Royal Arsenal at Woolwich took part in many of the trials of gunpowder and gun cotton, experimenting with suitable methods of firing both. Other experiments suggested by James Douglass included locating the charge at the centre of a parabolic reflector to enhance the directional quality of the sound but the 1½ lb charges used caused some damage to the reflectors! Even without reflectors, 1½ lbs of gun cotton made a sufficiently impressive sound signal to merit development of the electrical detonating devices.

By 1876, tests at Shoeburyness using slabs of damp gun cotton suspended from poles indicated the course of development of the signal, which was finally introduced into the Trinity House service.[10] After excess moisture was squeezed from the gun cotton, a six by six by one inch slab had a dry two ounce primer set into it, which was itself attached to 'a detonating tube and dynamo electric machine'.[11]

In October 1876, the signal was put into trial use at Heligoland.[12]

> *'Notice is hereby given that a Fog Signal will be established at about the middle of October; at Heligoland, which will, during foggy weather produce by the Explosions of Gun Cotton, a report similar to that of a gun, every fifteen minutes'.*

The keepers were instructed to continue firing '... until the fog or thick weather shall have entirely cleared away'.

Primers were coated in paraffin to resist moisture and were supplied in boxes of 50. The charges came in boxes of 48 or 23 depending on which of two types were used: a nine-ounce three-inch diameter charge about two inches long, or the 1½ lb slabs, six inches square and one inch thick. The detonators consisted of 20 grams of fulminate of mercury from which two lengths of insulated wire were attached, live and earth. At Heligoland, a large frame was constructed from pipework eight feet by seven feet high and the charge was suspended about halfway up. It was recommended that primers and charges be stored separately and if the charges dried out that they be carefully moistened and returned to the cool, dry, well-ventilated store. Instructions on how to prepare the charges and fire them were

---

[10] It was some time before all of the rock lighthouses were equipped with explosive fog signals after the first ones were placed in the mid 1880s. Godfrey Phillips in his 1949 book *Lighthouses and Lightships and the Men Who Man the Trinity House Service*, states that The Smalls was the site chosen for the first explosive fog signal but gives no date and he may have been referring to the gun cotton rockets used there to supplement the bell from 1880.

[11] *Notice to Mariners*, 29th September 1876.

[12] Heligoland, one of the North Frisian Islands, belonged to Britain after capture from Denmark in 1807. In a deal with Germany in 1890, it was 'exchanged' for Zanzibar.

supplied by Colonel Younghusband of the Royal Artillery. His early signals comprised the damp charge, a primer, a detonator and a length of twine with a hook and ring. The primer was pressed into the charge and the detonator pressed into the primer. With the hook and ring the charge was suspended on the frame and the wire from the detonator attached to the frame – 'earth'. The live line was attached to the detonator and the 'exploder' just prior to firing.

Colonel Younghusband was also the officer in charge of experiments with different ways of storing gun cotton. *The Engineer* in May 1873 reported on one of these tests, where two magazines containing wet and damp gun cotton were deliberately ignited. In neither case was there an explosion, but in both sufficient heat was produced by the slow 'flameless' fire to crack the magazine vaulting and melt the iron rails on which the cases of gun cotton were stacked!

The charges used at rock lighthouses were stored dry and there was a good safety record, but on two occasions there were serious accidents which caused injuries as well as damage to the two lighthouses involved. At Skerryvore on March 16th 1954, a fire started in the lighthouse, probably in the kitchen, and burned throughout the night, igniting charges and detonators. The keepers were safely taken off but the intensity of the fire caused extensive cracking and disintegration of the granite walls. Chicken Rock lighthouse also suffered from a severe fire which started during Christmas, 1960. On this occasion, it was difficult to get the keepers off the rocks by boat and impossible by helicopter; in addition to slight injuries, the keepers also suffered from hypothermia. As at Skerryvore, serious damage was caused to the building and it was not until September 1962 that work was completed on the refurbished tower. An electric fog signal was installed at that time and the lighthouse was made fully automatic. Explosive signals had been in use at Chicken Rock since 1885.

Far less serious damage occurred at Heligoland in the 1870s, where the firing frame had to be repaired several times. The frame had to be repositioned after initial damage was caused to the lantern and other windows at the lighthouse. The buildings, their contents and keepers all showed signs of exposure to the shock waves!

Although the Heligoland signal was dismissed by some sea captains as less audible than a 'fowling piece', the explosive signal nevertheless went on to provide an effective fog signal where the restrictions of space prevented a compressed air signal being used, and the same basic type survived for a surprisingly long time. Not until the 1960s and '70s were they finally replaced by modern compact signals and electric emitters.

Chapter 7

St Catherine's
and the 1901 fog signal trials

In 1900 the siren had been in use on British coasts for twenty-five years and there were over sixty in service.[1] The five-inch, cylindrical, automatic siren was in general use by Trinity House, with some sounding high and low notes. Many single-note sirens still sounded at relatively high frequencies.

A new siren design had recently been developed in Scotland. This siren was designed to sound both high and low notes but was motor driven up to speed before sounding, with the effect that the blast had a distinctive start in contrast to the automatic siren, with its rising note from a stationary start. In a similar way to the Irish siren, designed by John Wigham, compressed air rotated the turbine blades on the shaft of the Scottish siren to set it spinning. The seven-inch siren was attached to a new horn design which was horizontally elongated and had a wide mouth.

As the twentieth century approached, Trinity House had also been working on new developments. The Engineer in Chief was Thomas Matthews and the Scientific Advisor was Lord Rayleigh. Rayleigh had been experimenting with trumpet designs and Matthews had recently been reappraising disc and motor-driven sirens.

The most recent Trinity House siren installation was at Lundy North, which had the upright bell-mouthed trumpets as described in Chapter 4. These were made from cast iron and were twenty feet in length opening out to a copper bell-mouth, six feet from rim to rim. The trumpets stood vertically with their mouths bent through ninety degrees. Both trumpets contained a siren in the throat. This design replaced the earlier long horizontal horn, at lighthouses where fog signal equipment had been renewed and where a new signal had been installed, as at Lundy.

A similar fog signal with twin trumpets was in service at the lighthouse at St Catherine's Point on the Isle of Wight. The fog signal building was twelve years old

---

[1] In service at December 31st, 1901 there were (sirens, reeds, and hand horns respectively) England/Wales:32, 16, 29; Scotland:23, 4, 0; Ireland:10,2, 0.

Bell-mouthed iron trumpets as installed with the new fog sirens from 1890.

and situated near the cliffs, about eighty yards from both the lighthouse and the engine house. It was a single storey building with a hipped slate roof like other buildings at St Catherine's. Inside were receivers, clockwork actuating machinery, a recorder and the siren equipment. The trumpets protruded from the mid point in the roof on either side of the apex and faced approximately south-east by east and west by south, with the axes of the horns separated by 120 degrees.

The fog signal had twin, double-noted automatic five-inch sirens and the characteristic was two 2½ second blasts every minute. Compressed air for sounding the siren was pumped underground from the engine house.

The light at St Catherine's was an electric arc and much of the equipment had been installed in the late 1880s. Two De Meritens magneto machines, similar to those at the Lizard, were in use at the time[2] but at St Catherine's, the magnetos were driven by steam engines. The steam engines also worked the air compressors. The lighthouse tower at St Catherine's was built around 1839 but Walker's original design was later modified by the removal of nearly forty feet, in 1875. The octagonal castellated tower originally had three tiers and totalled 120 feet, but this height had made the lantern vulnerable to low cloud and fog. In 1868, the lighthouse had been fitted with a Daboll reed fog signal, complete with a rotating trumpet similar to the one at Dungeness.

---

[2]   The De Meritens magnetos were later replaced by Brush dynamos.

In 1887 Lord Rayleigh followed John Tyndall to the post of Professor of Natural Philosophy at the Royal Institution and he also succeeded Tyndall to the post of Scientific Advisor to Trinity House, although this was not until 1895. The post had remained unfilled since Tyndall's resignation in December 1884, following difficulties with James Douglass and the Elder Brethren.[3]

Rayleigh surpassed Tyndall by later becoming President of the Royal Society but, like Tyndall, had published a book on sound. The book was based on his own lectures on the subject, in which he had demonstrated various hypotheses in the laboratory using flames deflected by very high frequency sound, much as Tyndall had done. Rayleigh, however, was not convinced by all of Tyndall's views following the fog signal trials at South Foreland in 1873 and 1874 and was interested in conducting his own trials to further investigate the problems of silent areas and aerial echoes. Rayleigh was not alone in questioning Professor Tyndall's terminology and hypotheses. Professor Henry, who advised the US Lighthouse Service, believed that research into refraction of sound energy was likely to be more significant than the reflection of sound: Tyndall believed that 'aerial echoes' were proof of reflection from 'acoustic clouds'. Other scientists and engineers reopened the debate and some reverted to pre-Tyndall views on sound transmission in fog, believing that fog absorbs sound, a view encountered even today.

In one of his lectures, Tyndall asked his audience to visualize concentric 'shells' of air surrounding a sound-emitting device. For the sound energy to pass from one 'shell' to another, a greater number of air molecules would need to be agitated in the increasing area of the spheres and, in consequence, the sound energy would be proportionately diminished.[4]

By retaining the model of the concentric 'shells', we can visualize the spread of sound from a fog signal. Later scientists again picked up Professor Stokes' views

---

[3] A House of Commons review was set up into the merits of oil and gas for lighthouse illumination and a Commission was appointed by Joseph Chamberlain, the President of the Board of Trade. The report confirmed the superiority of the Douglass multi-wick oil-burning lantern, although Tyndall believed that the gas lighting system developed by the Irish engineer John Wigham was unsurpassed. Douglass subsequently patented his invention and set up his own manufacturing company much to the surprise of the Elder Brethren. A second major incident followed. The magnificent bi-form lenses such as those installed at Bishop Rock by Douglass had actually been developed by Wigham, who had registered the idea with the Patent Office. Douglass' actions led to legal struggles but Wigham settled out of court for £2,500. This money was paid by Trinity House, however, not Douglass! These matters underlined the souring of relations between Douglass and Tyndall. Long-standing differences between the men were aired in *The Times* in June 1883 and Chamberlain raised the affair in the Commons in 1884. Relations between Chamberlain and Tyndall deteriorated seriously. In the House of Lords, it was suggested that Tyndall might resign and Tyndall subsequently did so with 'inexpressible grief and pain'. Sir James Douglass survived unscathed and his biographer Thomas Williams barely referred to the incident!

[4] From Tyndall's book *Sound*, Lecture 1. 'Take the case of a shell of air of a certain thickness, with a radius of one foot, reckoned from the centre of the explosion. A shell of air of the same thickness, but of two feet radius would contain four times the quantity of matter; if its radius be three feet it will contain nine times the quantity of matter; if four feet it will contain sixteen times the quantity of matter and so on. Thus the quantity of matter set in motion augments as the square of the distance from the centre of the explosion. The intensity of the loudness of the sound diminishes in the same proportion.'

about air movement and the increasing velocity of air at increased heights. If we incorporate this theory, the concentric 'shells' would be distorted by wind speed and direction, so that instead of being spherical, the 'shells' would have elliptical cross-sections. The resulting effect of these elliptical 'shells' would be the deflection of sound energy upwards, possibly over the head of an observer up-wind, and a downwards deflection down-wind.[5]

By supplementing this model with further information on air temperature and humidity and by recognising that rising columns of warm air would also deflect sound upwards, and vice versa, it was thought that more light would be shed on the phenomena described by Tyndall and help to explain 'aerial echoes', silent areas, sound shadows and 'acoustic clouds'.

In the intervening years since the South Foreland trials, the demand for fog signals increased proportionally with trade. Although moving into an economic decline, Britain remained the major maritime trading nation with a total shipping tonnage which exceeded all other nations put together. There was still fierce competition between shipping companies; 'shipmasters too often ran on in foggy weather in the hope of picking up a sound signal' as the Trinity House Fog Signal Committee recognised.

It might be thought that a quarter of a century of research and development, including the installation and comparative evaluation of fog signals, would have resolved any outstanding uncertainties about the most suitable siren size, port size, port numbers, speed of rotation, air pressure, air volume, size and shape of trumpet and whether high or low notes were best!

By this time, some of the earlier sirens had been replaced and as the fog signals had been refurbished, oil engines were installed to replace the hot air engines. At most English fog signal stations, high pressure air was stored for immediate use when required. The performance of new equipment was, of course, closely observed, but it was felt that workshop experiments were insufficient to fully evaluate recent developments, and it was decided to undertake new comparative tests of siren equipment.

The site chosen for these new trials was St Catherine's lighthouse which had a large sea area and had sufficient capacity in the steam engines to provide plenty of stored compressed air. The service twin five-inch siren was to be used for comparison with all other sirens to evaluate effectiveness and economy of use.

Tests established that some retuning of the frequency of both high and low notes improved the effective range of the fog signals and the following frequencies were adopted for all five-inch sirens: 295 Hz for the high note and 182 Hz for the low note. The new Scottish seven-inch cylinder siren, and the experimental five-

---

[5]   The effect of the refraction of sound up-wind of a signal is as much a problem today as ever; moderate to strong winds can produce a shadow zone as close as half a mile from the signal.

inch and seven-inch disc sirens were to be compared with the service five-inch, and different horn types would be tested with all of the instruments.

In some respects the approach to the trials was the same as in 1873, as though little had been learned from the previous trials; similar questions would be asked and similar conclusions would be drawn.

However, the St. Catherine's trials also looked forwards as well as backwards. Although the same approach to testing air signals was to be laboured over again, in looking forwards, the Fog Signal Committee was open-minded about recent technological developments, anticipating new ways of safely navigating in fog.[6] At Knowles Farm, very close to the lighthouse at St Catherine's, Marconi had recently been undertaking experiments with radio and the potential application of radio[7] was recognised by the Committee even though '... conveying signals by electrical methods without introducing the element of sound ...' was beyond the scope of these particular trials.

Electric fog signals were not so very far into the future; bells could already be remotely controlled by electricity and electromagnets had been operating machinery for many years as at Ailsa Craig and elsewhere. Vibrating a reed by electromagnetism rather than compressed air had already been considered.

The trials were planned for May and June 1901 and in March, the Visiting Committee arrived aboard the *Irene* and noted the preparations. The Committee recommended that Blackwall staff be provided to help the keepers during the trials.

After the 1863 trials of the Daboll reed at Dungeness, five were installed followed by many more of the reeds designed by Professor Holmes. Even after Tyndall conclusively established the inferior properties of the reed, it continued to enjoy some success! After the Fog Signal Committee had organised the 1901 siren trials, the reed lobbyists again insisted on the inclusion of reeds.

Longstone ('Outer Fern') and Whitby lighthouses were shortly to be provided with new fog signals and the Elder Brethren wanted to evaluate a new type of reed then in use at Belfast. The reed fog signal remained effective at harbours and it was cheap to install and economic to run. If a reed was audible at two or three miles, what was the real advantage in a siren audible at twenty?

Four different types of reed horns were consequently included in the trials, two of which used multiple reeds. The Barker Horn had three reeds and was sounded at 60 psi, whilst the 'Stentor reed', supplied by the Pintsch Company, had a nickel

---

[6] Unpublished *Correspondence and Reports of the Trinity House Scientific Advisor* indicate that in June 1900 Rayleigh was interested in a proposal to make 'wireless electrical fog signals' but wary of patents, costs and reliability.

[7] Marconi was transmitting and receiving in Morse code but by 1906, it was possible to transmit music and voice.

steel reed and sounded at 120 psi.[8] The high pressure reeds had merits but were very expensive to operate. When they were sounded at the economically low pressure of 15 psi, the standard pressure at which Trinity House service reeds operated,[9] all merits were lost!

A fundamental problem with the reed signals was the extent to which they were sensitive to pressure variations in fitting and tightening the reed into position. This could affect the note produced and the output level. Any consistency required trial and error adjustments in the fitting and tightening. The reeds could not consistently compete with the sirens, even at the high pressures used, but the Committee felt obliged in their report to encourage further work by the Engineer in Chief, recognising that reeds still had a role. As we have already seen, reeds were subsequently used for many years and, following the St Catherine's trials, new reed signals were installed at rock lights such as the Wolf, in 1904 and the Needles, in 1906. At Hartland Point, a reed even replaced the siren in 1911.[10]

The trials ran from May 8th to June 13th in the presence of the Committee and with Lord Rayleigh, Thomas Matthews and the respective Engineers in Chief from the Northern Lighthouse Board and the Irish Lights, Mr D.A. Stevenson and Mr Scott, in periodic attendance. Other visitors, engineers, members of the Board of Trade and representatives of the Elder Brethren occasionally attended, but the repetitive nature of the trials was not of great interest to the casual observer.

There was a sense that theories of sound production and transmission would be weighed against empirical testing, as occurred at the South Foreland trials. As with the earlier trials, there were also occasions when the observers' proficiency would be challenged; observers sometimes heard signals that had ceased sounding or detected a qualified improvement when they were expected to.[11]

In a similar fashion to the earlier trials, observers recorded their responses to the different signals as heard from boats steaming up and down the coast. The difficulty of subjective assessment was mediated by recording the 'relative values' and subsequently deriving percentage figures to represent perceived differences. Over 4600 observations were recorded at sea and on land. In each case, the first sound heard was assigned the value 10 and the following sounds were then given greater or lesser relative values. Most observers appreciated the main formulations of the 1873 trial but there was still a sense of surprise and discovery, particularly in relation

---

[8]   The Stentor horn produced a note of 265 Hz. Pintsch was a German chemist who established a gas production company which produced equipment for buoys and lighthouses.

[9]   The Trinity House service reed produced a note of 397 Hz.

[10]   The reed replaced an early two-note siren in January 1911 when new oil engines replaced the old caloric ones. 'The spare reeds are stored in a suitable place in oiled paper' says the station Order Book. The reed signal gave way to a new diaphone in 1927.

[11]   At a trial of lighting equipment, observers aboard *Irene* noted an improvement in the brightness of the light although it later transpired that the new apparatus had not been used because of a mechanical problem!

to the observed silent area and the duration of echoes on fine days. Even Rayleigh was surprised by his experience of the aerial echoes which Tyndall had described.

The recognition of the effect of varying densities of air and wind on sound transmission did not spoil the attempted objectivity of tests with different equipment on different days. The report on the tests did concede, however, that because the experiments were '... not being made under equal conditions ... the result ... was somewhat uncertain'.

As with the 1873 trials, variable conditions produced variable results – under some conditions a high note was superior to a low note and vice versa, but how could this translate into workable policies at fog stations? Variations in the pressure at which sirens were sounded made little difference in most circumstances, although the higher pressures produced marginal improvements.

The tests proceeded with four siren types,but after trying them in different meteorological conditions with different trumpets, the daily results varied and Rayleigh observed, with disappointment, that there was '... very little difference between any of the things we have tried yet'.

Sirens tested in 1901 trials.

Siren type	Note		Number of ports		Amount of air used (cu. ft/sec)	
	high	low	high	low	high	low
5″ experimental	295	182	24	16	32	16
5″ disc	250		30		16	
7″ disc		98		18		36
7″ cylinder	234	100	14	6	130	26

The seven-inch disc siren was superior to all others except when used with the mushroom horn,[12] when it was lost before the five-inch service siren. The Scottish trumpet was tried with the service siren and the seven-inch siren; the Scottish siren was also tried with the mushroom horn. The motor-driven sirens were marginally better under most conditions but all of them used much more air than the service five inch.

---

[12] The so called 'mushroom horn' was similar to those described at St Catherine's and the Lizard in overall dimensions except that the bell mouth was not bent forward through ninety degrees but pointed straight upwards. A deflector was attached to the trumpet rim, which was designed to prevent sound energy being wasted vertically and to reflect sound through 360 degrees of arc. It was generally intended for use on light vessels but there were lighthouses where there was space for a siren and a very wide sea area to protect.

Although slightly different in the notes produced, the performance of the seven-inch cylinder and the seven-inch disc could not be substantially differentiated by observers. Further development of the discs was agreed, but aboard the *Irene*, as Rayleigh compared the five-inch disc with the five-inch service cylinder, he could only admit that the '... whoop of the old signal proved to be the best part of the whole and was heard at ten miles when all else was lost!'.

It was noted that the disc sirens worked better when 'tuned' to the horn being used and the 'Caskets' seven-inch disc was enhanced this way, the 98 Hz note being lowered to 93 Hz. Tuning sirens and reeds to the trumpet was established practice, though of less significance with the cast-iron or wrought-iron trumpets. The performance of copper and brass trumpets was definitely improved by tuning the siren, as the French had discovered.[13] French thinking also regarded larger trumpets as wasteful of sound energy and that their use produced a distorted note.

The report on the trials recommended tuning sirens to their trumpets and Corporation steam tenders were thenceforth to be furnished with the necessary equipment to check and tune sirens. A siren should be tuned to the fundamental note of the trumpet, and to gain most benefit from the high and low note sirens, the two notes should be separated by an octave. An alternative idea was to design the horns to accord with specifications prescribed by the siren.

Lord Rayleigh was responsible for new developments in the design of fog signal trumpets, developing the relationship between the wavelength of the sound and the dimensions of the horn.

One of the aims of the trials at St Catherine's was to determine the most effective way of protecting a large sea area, but the merits of the new sirens were still undermined by the directional horn; the sound level was diminished off-axis. The twin siren consequently fared well in many of the trials because of the 120-degree spread between the trumpet axes.

For a directional trumpet, Rayleigh calculated that the diameter of the mouth should be a multiple of the wavelength of the sound produced. For a wide distribution of sound, up to 180 degrees, his theory was that

> '... the diameter of the aperture must not exceed the half wavelength [of the siren note] otherwise there will be serious interference between the parts of the sound proceeding from the various parts of the aperture'.[14]

---

[13]   The French trumpets were brass, four feet six inches long with a maximum diameter at the mouth of three feet one and a half inches. Siren trumpets were placed high on the lantern galleries and French tests had established the optimum size, believing that larger trumpets caused sound distortion and inferior quality. After the St Catherine's trials the French trumpet was subsequently included in further tests at Souter Point later in the year and at the Trevose trials in 1913.

[14]   Strutt, Robert John, Fourth Baron Rayleigh FRS, *The Life of John William Strutt. Third Baron Rayleigh, OM FRS*, 1924.

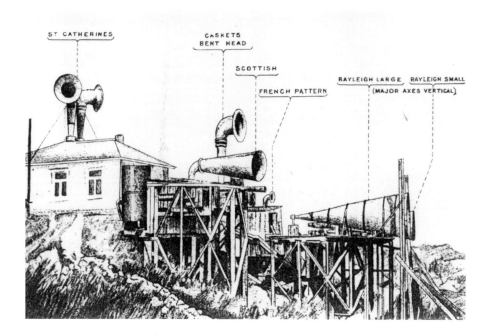

Different trumpet designs used in the 1901 trials. Courtesy of Trinity House.

In essence, theory suggested that sound waves could be transmitted in opposite phase and, if so, this could result in a diminution in amplitude. Rayleigh further believed that a circular mouth wasted sound energy since the design could not maximise the horizontal spread of sound and he designed an elliptical trumpet, calculated on the principle of the half wavelength. The actual dimensions of the horn would depend on the pitch of the note sounded; for example, a sound of 256 Hz, the 'high note', produced a wavelength of 4 feet 4 inches, whereas the 'low note' of 182 Hz had a wavelength of 6 feet 2 inches.

Large and small Rayleigh trumpets were designed and used in the 1901 trials at St Catherine's. The larger trumpet was made of wrought iron and the elliptical cone was 20 feet long and tapered to a 5-inch diameter at the throat where the siren was attached. The elliptical mouth measured 7 feet 4¾ inches by 3 feet 7¾ inches. The small trumpet was also 20 feet in length with a 5-inch throat, but opened out to a vertical axis of 4 feet 10 inches with a horizontal axis of 1 foot 1¾ inches.

The Rayleigh trumpets, despite the theory, did not show a marked improvement on the service trumpets. In six trials on 28th May, on only two occasions did the Rayleigh trumpet achieve better results than the twin trumpets, and these were both on the medial line between two and four miles from the lighthouse. Off-axis, the service twin trumpets were better each time.

On a warm calm day when tested against the Scottish siren, the larger Rayleigh trumpet was 14% better, both on and off-axis. On this occasion, the siren in the Rayleigh trumpet was the seven-inch disc rather than the five-inch service siren. Based on this notable improvement further trials were recommended, although the improvement seems to question the theory behind the trumpet's proportions relative to the wavelength!

At the conclusion of all of the trials at St Catherine's, the report produced by the Fog Signal Committee was inconclusive about Rayleigh's trumpet design[15] and suggested that

> '... *further investigation should be undertaken by Lord Rayleigh and the Engineer in Chief, with a view to the determination of the question as to its suitability for installation at a fog signal station ... the Committee are not ... prepared to make any definite recommendations ... in relation to the Rayleigh trumpets'*.

The trumpet design was nevertheless installed at the new fog signal station at Whitby[16] within eighteen months.

In 1903 Rayleigh demonstrated a variation on the trumpet design to the Elder Brethren and Thomas Matthews. This instrument seemed to prove Rayleigh's theory without question and a scaled-up version of the device was finally constructed at Trevose Head lighthouse in Cornwall. In the meantime, the oval trumpets were introduced at several new fog stations and as replacements at two others.

In April 1903, the first of the new fog stations was established at Whitby. Rayleigh's intention was for a single trumpet to cover 180 degrees of arc but the success of the St Catherine's service trumpets in the trials resulted in the following arrangement being placed at Whitby: two of the eliptical Rayleigh trumpets, each with a five inch siren, were positioned on the roof of the fog signal building with their axes separated by about 120 degrees. A Notice to Mariners issued at the time encouraged comments from mariners on the merits of the new signal and within a short time, Nash lighthouse in South Wales had a similar building and trumpets. Flatholm, in the Bristol Channel, had similar trumpets and sirens in 1907, as did

---

[15] The main conclusions drawn from the trials were that: 1. A wide sea area required two horns for adequate protection, although 'the mushroom horn was satisfactory'. 2. Motor-driven sirens were better than the automatic. 3. Twenty-five psi was the optimum pressure. 4. Sirens and reeds should be tuned to the fundamental frequency of the trumpet or trumpets. 5. High and low notes should be separated by an octave. 6. Under calm conditions, with the wind or across the wind, the low notes were best. But against the wind or a noisy sea, the high notes were better. As a consequence of this recommendation, the Committee suggested retaining high and low note signals in service for the time being. 7. A ship should be stopped and all noise aboard quietened in the vicinity of a fog signal. 8.'... A fog signal must be regarded only as an auxiliary aid to navigation ...'. 9. Two and a half seconds was the shortest blast to be used and five seconds the longest, but because of the likely interference with navigational signals between shipping, the otherwise unnecessary length of seven seconds was to be used. 10. It was noted that a long signal duration had a stronger impression on the observers.

[16] Further to the St Catherine's trials, both Whitby and Longstone had sirens and not reeds. Whitby was fitted with the new Rayleigh trumpets, but Longstone had the older upright bell-mouthed trumpets.

Flatholm fog signal with its twin Rayleigh trumpets, circa 1920. Courtesy Trinity House National Lighthouse Centre.

South Bishop lighthouse. At a later stage, the existing fog signal buildings at Souter Point, Flamborough Head and Bull Point were also fitted with similar equipment,[17] but Alderney and Round Island were the last new stations to be equipped with the Rayleigh trumpets, in 1912.

The new fog stations all had a similar layout, except at South Bishop and Round Island (due to space) and Alderney, where the fog signal building was attached to the front of the tower. In the others, the building was separated from the lighthouse by some distance. The fog signal building had three rooms, two small ones for fuel storage and a workshop with a larger main engine room, which contained two 22 hp Hornsby oil engines.[18]

A Hornsby single cylinder compressor, 11¾ by 14 inches, was driven directly by the engine and supplied two large storage receivers which fed into a smaller sounding receiver situated between the larger tanks. Above the receivers was a roof turret which contained the two sirens. The large receivers were 14 feet 6 inches high and each contained 265 cubic feet of air at 150 psi. The sounding receiver had a capacity of 45 cubic feet and was linked to the larger tanks via a reducing valve set at 25 psi. Chains hung from the clockwork actuating machinery high up on the wall. A ladder beside the small receiver led up to the siren turret within which was a large, cast

---

[17]   Flamborough Head had Rayleigh trumpets installed in 1908. Souter Point and Bull Point both had Rayleigh trumpets in 1919.

[18]   The Whitby and Nash Hornsby engines were bought in April 1903 and had consecutive serial numbers.

Nash engine room with Hornsby. Courtesy of Trinity House.

Pedestal and twin sirens inside the turret at Flatholm.

iron standard holding the two siren chests up to the throats of the trumpets, along with the geared or belted regulator and the six-inch admission valve. The turret was seven feet in diameter and five feet high, with the centre line of the trumpets three feet above the roof of the fog signal building.

The trumpets were eighteen feet six inches long and made of wrought iron with cast-iron ribs and steel mounts. The vertical axis of the oval mouth was seven feet six inches and the horizontal axis was two feet nine inches. The axes of the two trumpets were separated to provide the greatest area of cover. At Alderney, the angle between the trumpets is 120 degrees but on the island of Flatholm, the angle between the trumpets reflected the greater cover required and the trumpets were spread by 150 degrees.

The sirens were manufactured by Chance Brothers and all sounded 'low' notes, 180 Hz. Except for Round Island, the character of the other Rayleigh signals was sounded every ninety seconds. Whitby, Alderney and Nash all sounded four blasts.

The Hornsby engines were in use at all the Rayleigh stations until they were replaced in the late 1960s and the early 1970s, usually by three- or four-cylinder Gardners and Reavell compressors. The trumpets were refurbished with mild steel in the 1950s and 1960s. The Rayleigh trumpets were also used at other stations, either as replacements (sometimes temporary) or for other trials. At Souter Point, evidence of the Rayleigh trumpets is still visible inside the turret, where the oval apertures have been welded over.

The improved prototype trumpet, which Rayleigh demonstrated in 1903, developed the principle of the elliptical horn with its dimensions proportional to the wavelength of the specific sound for which it was designed. The scaled down prototype used a small reed which sounded at two octaves above the service reed horn, with a wavelength of approximately eight inches. Rayleigh constructed a wooden pyramidal trumpet six feet long with a rectangular mouth thirty-six inches high and four inches wide.

As with the oval trumpets in the St Catherine's trials, the vertical and horizontal axes were altered between demonstrations so the observers could gauge the effect. Where popular belief would expect the sound to be better spread with the major axis horizontal, this proved not to be the case as Rayleigh had calculated. With the major axis vertical and the observers positioned thirty degrees off-axis, the performance of the horn was considerably enhanced as theory predicted.

A full scale version would need to proportionally increase the dimensions of the horn in relation to the longer wavelengths of the lower note and in 1911 such a device was built on to the newly-constructed fog signal building at Trevose Head. The scaled up trumpet was thirty-six feet long, two feet wide and eighteen feet tall at the mouth. Rayleigh designed the new horn

> '... for a pitch of 256 vibrations per second ... if anticipations are realised, the sound should spread pretty equally over an angle of 180 degrees, ninety degrees each way from the axis'.[19]

He did not believe the construction material was particularly significant as long as the interior surface was

'... unyielding ... and airtight'; Rayleigh believed that concrete, iron or brick with oval or rectangular mouths would all have produced the same effect.

---

[19] From *Correspondence and Reports of the Trinity House Scientific Advisor*, letter dated 28th September 1910.

The unique Rayleigh trumpet at Trevose Head. Courtesy of Trinity House National Lighthouse Centre.

On completion, the trumpet, which was finally made of timber with iron supports set in concrete on the clifftop, looked similar to a huge trumpet built by Daboll for a reed tested in America forty years previously.

Trials of the new trumpet began in May 1913, when Matthews and Rayleigh steamed to various co-ordinates off the headland at Trevose listening to the new trumpet in comparison with a French horn (the same one used in the St Catherine's trials). Once every minute throughout the day one or the other horn was sounded. In the Rayleigh trumpet, a service five-inch siren was installed (180 Hz) and this was alternated with a 'new small port siren running at 440 revolutions'. The seven-inch, motor-driven disc siren, which had been tested at St Catherine's, was now fitted to the Rayleigh trumpet and run at 700 rpm. It used twice as much air as the service five inch, but with the instantaneous start was expected to be impressive. Sadly, the anticipated results were not achieved and Rayleigh ascribed this to the failure of the siren to spin at the correct speed of 960 rpm which was required to produce the prescribed pitch of 256 Hz. The maximum speed obtainable was 750 rpm which would have produced a note of 200 Hz.[20] In a letter to the Elder Brethren,

[20] Sir John Bowen was the Engineer in Chief to Trinity House from 1924 to 1951 and in his 1947 book, *British Lighthouses*, he states that Rayleigh calculated the proportions of the Trevose trumpet based on a wavelength of six feet, so that the length, thirty-six feet, is six wavelengths and the height, eighteen feet, is six half wavelengths. A six foot wavelength is produced by a note of 182 Hz: the service five inch siren. However, Rayleigh did not design the trumpet for a pitch of 182, but 256. As noted, the engines could only spin the siren at 750 revs per minute, so that the siren would have produced a note of 200 Hz and, interestingly, the proportions of the trumpet in relation to this wavelength would again come nearer to Bowen's figures! The wavelength of 200 Hz sound is five foot eight.

Rayleigh thought that '... a new engine to drive the siren would be required. Probably the Elder Brethren will not wish to embark upon this at present ...'. Probably not, since the building and engines at Trevose were brand new!

After the trials of the new horn, the observers concluded that a similar single horn proposed for installation at Souter Point should be deferred since '... the interests of the navigators would be better served by two horns of the older Rayleigh design'. These two horns were installed in 1919.

At Trevose Head, the service five-inch siren was retained in the huge Rayleigh trumpet and this unique monster remained in service until 1963 when it was dismantled. Eight supertyfon air horns replaced the Rayleigh trumpet and new engine and compressors replaced the originals.[21]

Rayleigh continued to experiment with sound transmission and reception[22] but his experimentation with fog signal equipment ended at Trevose. He remained as Scientific Advisor but Trinity House did not seek to appoint a successor after his death in 1919.

After the 1901 fog signal trials, the five-inch siren and twin trumpets remained in service at St Catherine's, but by the late 1920s the fog signal building had serious cracks throughout, caused by the collapsing cliff. A new siren was installed in 1932 fitted in a cast-iron turret on top of the newly-constructed fog signal tower built in front of the lighthouse tower.

The lighthouse was attacked during the Second World War and three keepers were killed when the engine house was destroyed by a bomb. In the major refurbishment of the lighthouse after the War another fog signal was installed, a diaphone. In 1962 this signal was replaced by a two-tone supertyfon which produced a distinctive chord every forty-five seconds during fog and remained in use for over twenty years.

Despite its history, there is no longer any kind of fog signal at St Catherine's.[23]

[21] Over the years, the unrelenting cliff erosion produced large cracks in the walls of the fog signal building and when the station was automated in 1995, the building was demolished.

[22] Rayleigh devoted some time during World War One to developing 'acoustic eyes', parabolic reflector dishes designed to aid the detection of enemy aircraft by concentrating the sounds picked up. Similar equipment was installed near Dungeness in the Second World War, prior to the development of radar detection, which rendered the facility obsolete.

[23] St Catherine's lighthouse became the main centre for experimental trials of various types of navigational equipment when Blackwall workshops closed and the Engineering Department was re-organised and relocated – the Research Department moved to East Cowes.

St Catherine's lighthouse and fog signal tower.

Chapter 8

Lundy

Refurbishment and new plant

At low tide the channels of the river Torridge are insufficiently deep for the MS *Oldenburg* to navigate, but when the tides are right the vessel usually sails from Bideford to Lundy several times each week. On a bitterly cold October day the boat is now moving slowly towards Crow Point light in the widening channel. Once beyond the sand bars we are rolling pleasurably, if not a little excessively, into Barnstaple Bay and the vessel aims towards the angular grey block on the horizon, fully exposed to the easterly wind. The sea is a green variation of grey and some of the wave tops tear open. Opening to the left though merging with the cliffs, Hartland Point lighthouse blinks automatically and shortly coming into view on the right will be Bull Point.

As Lundy looms up, the prominent church tower and old lighthouse on the horizon are joined by the collection of white buildings on the south-east headland. The landing bay affords shelter from prevailing south-westerly winds and seas and provides the only easy landing on the island. In strong easterly winds, however, the normally straightforward process can become difficult.

The *Oldenburg* drops anchor and swings to face the waves as the small boat from the island laboriously braves the seas to collect the passengers. Both vessels roll as the passengers are awestruck by the sheer cliffs, wondering which is the lesser of two evils. Should it be staying aboard and hoping the seas calm down on the return to Bideford or stepping off the *Oldenburg* into the landing boat, wondering about the depth and temperature of the sea and whether the waves are getting bigger? Will the waves break directly over the little vessel or simply destabilise it by washing it against the side of the *Oldenburg*? Weighted down with luggage, neither the boat nor its occupants seem very buoyant, though an artificially buoyant mood spreads amongst the fearless and the fearful!

Ashore at last, the relief is measured by hyperactivity; the imagined headlines reporting a major boat disaster are now lightheartedly aired! The seas were large but fortunately, the little boat rose up on the wave rather than dipped into it.

However, this book is not about shipwrecks but the means of preventing them and as Charles Ingrey proclaimed in 1886: 'The siren … has been the means of saving many wrecks and preserving thousands of lives'. Nevertheless there have been shipwrecks at most of the locations described in this book, sometimes in good conditions and sometimes within feet of the lighthouse. Around Lundy's coastline are the wrecks of 137 ships, one of which occurred in the early hours of 30th May 1906.

HMS *Montagu* was undertaking a clandestine operation of some sort, but possibly only experiments with radio, and was heading up the Bristol Channel in thick fog. Evidence from the local lighthouses established that all of the fog signals were operating that night and judging by the evidence Captain Adair presented at his court martial, he heard all of them. Unfortunately, the captain believed his ship nearer Hartland Point than Lundy and, accordingly, ripped a hole in the hull of the *Montagu* and lost one of his crew before realising his error. Having heard what he described as a siren sounding 'two high and two low notes',[1] the Captain convinced himself that the sounds were from a steamer.

The siren at Lundy North sounded low, high, low, high every two minutes and the siren at Hartland Point sounded 'Two quick blasts, high and low, every two minutes',[2] as the Admiralty List of Lights described it. Captain Adair believed HMS *Montagu* was making for Lundy from the south and expected to hear the explosive signal from Lundy South lighthouse. He heard another siren which sounded a long single blast.

At the south-west point of Lundy there are reefs which stretch some way out to sea, and it was on these that the *Montagu* struck. A serious problem became catastrophic as the vessel, with only one propeller, could not easily free itself and it finally grounded below the 100 ft cliffs near Shutter Point.

Shutter Point is in a potential sound shadow of Lundy South signal and even in clear visibility, approaching the island on this course, the headlands can screen the lights of both lighthouses close to the island. The concluding report of the 1901 Fog Signal Committee asserted that '… a fog signal must be regarded only as an auxiliary aid to navigation …'. It was still unknown if silent areas occurred in foggy conditions and it was, in any case, recognised that there could be no guarantee of hearing a signal at a specific distance. For many years captains had been urged not to use the apparent loudness of a signal as any kind of indication of proximity to it. In contrast to the silent areas were the aerial echoes which Tyndall had discovered

---

[1] Davis, G.M. & R.C., *Trial of Error*, 1983.

[2] It was usual practice to inspect fog signal records and check the equipment after such an accident. There is no indication that the fog signal at Hartland Point was not operating normally at the time, although the siren had previously caused problems. The Order Book indicates that '… only one of the two notes which form its proper character…' was sounding in 1900, with the siren '… usually missing the low notes'. The siren was replaced by a reed fog signal.

and which were also described at St Catherine's. The duration of the echoes heard near the fog signal, were inversely proportional to the strength of the sound blast recorded by observers at sea.

Echoes heard at St Catherine's had sometimes lasted up to ten times the duration of the blast and seemed to fill the air so that it was impossible to identify the direction from which the sound came. The two-note siren was originally designed to be clearly distinguishable from other signals and the characteristic assigned to a fog signal had to account for the navigational signals conventionally used aboard ships, not least the four to six second blast every two minutes, which ships were required to sound in foggy conditions. At the turn of the century, the performance of the two-note siren was causing some concern. It was suggested that the high and low note might actually cause problems rather than assist identification, especially in conditions where the ranges of the separate high and low notes were inconsistent, or where one frequency might even be lost altogether. The St Catherine's trials showed that low notes were usually reliable with or across the wind, although the high note was more successfully heard against the wind and noisy seas.

It seems very likely that some of the conditions described above played a part in the loss of the *Montagu*. Had Trinity House officials been represented at the court martial, some of the above information may have been presented for consideration.

Adair should have been sounding his own fog signal and he should have also listened for its echo from the cliffs as he approached Lundy. It was customary to stop engines and listen for a fog signal from the quietest parts of the ship. It is not clear why the usually reliable explosive fog signal at Lundy South was not heard aboard the *Montagu*. From Adair's description, it would seem that against the wind and over the sound of the surf and the shipboard sounds of the *Montagu* it was not possible to hear Lundy North siren sufficiently to correctly identify it. However, he could not have been unaware of the possibility of picking up Lundy North as his charts would have distinctly shown the lighthouse and the fog signal characteristic. There is no doubt that the transmission of sounds can be unreliable and the distinctive characteristic may have been further distorted by echoes close to the cliff. I have heard the two distinct blasts of a signal become slurred by echoes into one long blast on several occasions near different lighthouses.

The area which the twin trumpets of Lundy North were intended to protect was very wide, nearly 290 degrees of arc, but the low, high, low, high blasts every two minutes had been in successful operation for nearly ten years by this time. The two lighthouses at the north and south of Lundy came into service in 1897 when the original lighthouse was discontinued and, at the time, Lundy North was regarded as the '... most scientific and most advanced lighthouse in the world'.[3]

The lighthouse was built on a cleared site at the top of a 150-foot cliff, but behind the lighthouse, steep rocks climb as much again before reaching the flat top

[3] *Engineering*, 7th January 1898.

of Lundy, a gently undulating plateau elevated by sheer cliffs to over 300 feet above sea level. Supplies for constructing the light were facilitated by the construction of a small quay built in the granite cliffs linked by steps and 100 yards of narrow-gauge rails, which are still present. The lighthouse buildings are split on three levels, linked by steps and corridors with the small tower in the middle and the engine room and fog signal nearest the cliff edge. Matthews designed the lighthouse and with Rayleigh, introduced the first mercury trough at a British lighthouse. Floating the 3½ ton optic on mercury enabled quick and easy rotation and the four-panel optic produced a group flash of two every 20 seconds. The French optic[4] was rotated by clockwork drive regulated by George Slight's governor, a variation of the type used for regulating sirens.

In the engine room were two Hornsby engines with five foot diameter flywheels, which ran at 200 rpm and produced 16 bhp. Manlove Alliot compressing pumps were directly driven from the crank shaft and underfloor pipework conveyed air to the storage receivers. The engines were cooled by rainwater collected on the roof of the engine house and stored in a 7000-gallon tank on the cliff beside the engine room. Hot water, returned from the engines, poured on to the surface of the tanks and cooled before over-flowing back inside.

The engine room was thirty feet wide and extended twenty feet from the door to the siren equipment, which was situated in a characteristically curved bay in the front wall. From the door, the two engine sets were on the right. The bed-plates were nine feet by seven feet, leaving about a four foot space around the engines. To the left of the entrance a door led into the small workshop and the two air receivers, each with a capacity of 175 cubic feet, were situated between the end wall of the workshop and the far wall of the engine room to the left of the siren. This layout is very similar to that at Pendeen lighthouse, built three years later. One engine was used for compressing air with one on standby, and the receivers were both filled with air at 170 psi so that the signal was ready for immediate use. Stored air could supply the siren for about half an hour via the reducing valve fitted to the sounding receivers situated below the siren.

The two-note siren had two admission valves on two separate sounding receivers. The alternating low and high notes were produced when the valves were opened by the actuating clock on the wall beside the siren. The sirens were the familiar five-inch, cylindrical, automatic, two-note instruments and the trumpets were the vertical types with bell mouths set with their axes at 180 degrees to cover the wide sea area to be protected.

The lighthouse at the south of Lundy was built on a larger plateau, but at a similar height above sea level. The layout of the buildings reflected the larger site and included a magazine with its barrel vaulted roof, which stored the explosives

[4] James Douglass favoured Chance Brothers but believed Barbier and Benard of Paris made superior lenses – '… the arrangements for testing the optical adjustments … (were) … better'. After Lundy, St Mary's, 1898, was also French but Pendeen, 1900, and Beachy Head, 1901, were Chance Brothers lenses.

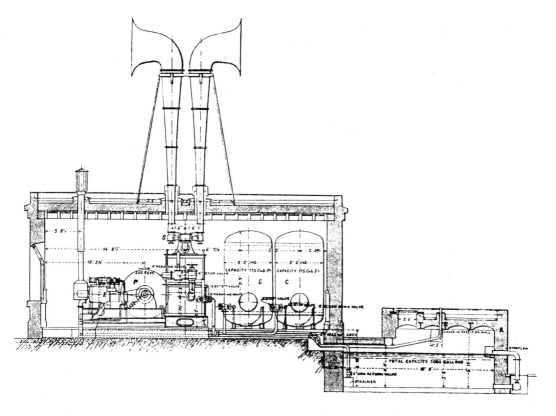

The 1897 fog signal machinery at Lundy North. Courtesy *Engineering.*

used for the fog signal. The original tubular steel, explosive jib attached to the lantern gallery was replaced by an improved explosive signal about 1908. Inside a small building made of riveted iron plate was an extraordinary device, manufactured in Victoria, London, by the Clockwork Explosive Fog Signal Company. Extrapolating on the name of the company I am sure you are already ahead of me!

Just off centre in the small building was a large drum, about two feet six inches in diameter. Around the circumference of the drum were over forty lengths of tube. Above and to the side was a clockwork mechanism driven by weights falling in a column and this column protruded through the roof of the building to support the transport system. Within a protective metal sleeve, primed tonite charges, as used on the jibs at rock lighthouses, were placed in the tubes around the drum. The clockwork rotated the drum and powered the chain-driven transport and in geared synchronization, the hook on the downward moving part of the chain engaged a charge by its sleeve and lifted it out of the tube and up and out of the aperture in the roof. As the charge passed out of the roof, the angle of the chain drive enabled the charge to be drawn from the sleeve. The charge continued to be conveyed some

*Clockwork explosive fog signal
similar to that used at Lundy South.*
Courtesy Trinity House.

fifteen feet clear, before making electrical contacts with the battery inside the
building, at which point the charge was fired. A safety feature transported unexploded
charges into a tank of water. Sadly, on one occasion it must have failed to do so, or
some other catastrophe occurred, because the iron building and its contents
exploded! A similar device, but made by a different manufacturer, was in service at
Lynmouth until 1958. The iron building at Lundy was located on the site now
occupied by the helipad. The explosive fog signal was succeeded by supertyfon air
horns in 1964.

The island of Lundy is a major obstacle in the route of shipping in and out of
the Bristol Channel and before the new lights came into service, Lundy had been
provided for by an impressive stone tower built by Daniel Alexander in 1820. A
light was shone at nearly 400 feet above sea level, making it one of the highest in
Britain, surpassed only by Cape Wrath and Barra Head, but low cloud proved to be
a major problem. The lighthouse had an early type of rotating apparatus which
completed one revolution every sixteen minutes, although it produced a flash every
two minutes.[5]

---

[5] An early revolving light was installed at St Agnes lighthouse on the Scillies about 1790. A three-sided
framework had seven lamps on each side. Lundy had similar reflectors and Argand lamps.

A secondary light, originally placed halfway up the tower, was designed to be visible over a projecting lip to vessels closer than four miles from shore, but from a distance, the two lights appeared to merge. When the secondary light was placed in a lamp house at the base of the tower there was little more success.

The low cloud and fog led to the light being augmented by a cannon, originally fired near the lighthouse from 1861, but in 1863, a special battery was constructed on the western side of the island. A steep, walled pathway leads down the precipitous slopes and cliffs to the small derelict cottage in which the fog gunners lived with their families. Still pointing out to sea are two guns and their carriages, on either side of the battery building. Crouching in the cliffs are two other magazine buildings and the gunners' cottages.

The origin of the guns on site is unclear, but they date from George IV's time and were, therefore, presumably made between 1820 and 1830. From 1863, two eighteen-pound cannons were fired from the battery in foggy weather every fifteen minutes at an annual cost of £350. The accommodation, ancillary buildings and cannon cost £1396 to construct.

The guns were muzzle-loaded and took some time to prepare for each firing. The two gunners could not reload within fifteen minutes, even though Trinity House recognised that a five minute interval was more desirable. The guns fired a three-pound charge of powder in an 'ordinary flannel cartridge, with a wad made of coir junk'.[6] There were over 500 hours of fog during 1873, during which the guns were fired 2065 times.

The firing of guns had a long history at sea and, subsequently, at naval stations. There may well be unrecorded instances of guns used for fog warnings at other coastal stations, but in America the practice dates from 1720 at Boston. The earliest documented use in England is from 1777, at Bamburgh, where a gun was fired every fifteen minutes. The Trinity House stations at Flamborough Head and North Stack fired eighteen-pound cannon, as at Lundy. The Flamborough guns dated from 1861, the same year that Lundy introduced the fog gun, but at North Stack the site was acquired from the Admiralty. The List of Lights from 1865 reports that 'a gun is … fired from the North Stack every hour and half hour (Greenwich Time) during foggy water …'.

There was still much support for the use of guns as signals after the South Foreland trials and Tyndall was obliged to undertake further experiments. At South Foreland the guns had been heard for many miles on occasions, but their drawback remained the brief period of sound and the long interval between firings. The brief sound was susceptible to being lost in strong winds and the nature of the sound, in combination with the long intervals, made the signal difficult for a mariner to attentively listen out for. To convince eminent members of the Admiralty and the Elder Brethren of the shortcomings of guns required many more tests.

---

[6] Letter from Trinity House to the War Office, 28th February 1874.

The battery at Lundy.

After Tyndall's report on the South Foreland trials was circulated, the Assistant Superintendent of the Royal Gun Factories, Major Maitland, distributed his own paper asserting that the report of the gun was distinct and better able to be recognised by any mariner than the prolonged note of the siren! He thought that the siren might be used on lightvessels but that guns should be used ashore. Of the guns tested at South Foreland, the 5½-inch Howitzer usually fared best. The report from the Howitzer was distinct, whereas the eighteen-pound cannon often produced a sustained rumble. Trials with gunpowder at Woolwich Arsenal established that fine grain powder was the most effective for producing a loud bang, although it was usual practice to use coarser grained powders with larger guns. While 'pebble grain', as it was known, satisfied the objectives of explosively hurtling a projectile from a large gun, the large grains produced a slow rumble. Massive guns of eighty tons using 400 lbs of powder had been fired, but these merely sounded like distant thunder. The Woolwich tests used 4½ lbs of powder in a 24 lb Howitzer and produced reliable, distinct bangs which carried over two miles; with a 6 lb charge it was felt that they might be trustworthy up to five miles.

As noted in chapter 6, Professor Tyndall was familiar with Frederick Abel's work which prevented gun cotton from spontaneously exploding. Trials with gun cotton consistently produced sharp reports which were superior to the guns. However, Tyndall also prudently recognised that the cost of gun cotton was three times greater than powder!

Further tests with gun cotton moved downriver from Woolwich to Shoeburyness and Mucking Flats and thence further afield. In 1877, at the Shambles lightvessel off Portland Bill, the service siren was compared with a twelve pound carronade with a one pound charge and in the same year, Mr Mackie of the Cotton Powder Company lobbied for his products to be included in tests.

During the experiments with gun cotton, Admiral Sir Richard Collinson, the Deputy Master of Trinity House, suggested that no acoustic shadow would interfere with the spread of an explosive report if the charge could be detonated at a height above the station, perhaps by the use of a rocket. The sound would be spread in all directions. Mr Brock, 'the pyrotechnist', was invited to assist with developments and in July 1876, eight rockets were demonstrated at his firework factory. Four rockets had 5 oz charges and four had 7½ oz charges of gun cotton. The results were impressive and further trials followed at various lighthouses over the following months. In August the following year, trials were conducted at Lundy where rockets were compared with the eighteen-pound cannon. Observations were made from positions open to the battery and others screened by the land itself, but it transpired that the gun was consistently inferior to the rockets, even when the observers were out at sea on the west coast.

Further trials took place at Flamborough Head later in the year and after due deliberation, rockets replaced the cannon at Flamborough from January 1878. Rockets were in regular use at Lundy in 1879. Each rocket cost one shilling and seven pence ha'penny and the Trinity House Secretary calculated the cost of firing them at intervals of two minutes, five minutes and ten minutes for cost effectiveness.

The interval between rockets fired at Lundy was ten minutes. A metal framework was installed beside the battery wall for launching the rockets. Before firing, the rocket had to be fitted with a detonator – a percussion cap filled with fulminate of mercury – which was inserted inside the explosive charge. After reassembling, a six-foot long stick was also attached to the rocket which was then placed in the firing frame. Launching the rocket was simple enough and at a height of 600 feet the propellant was exhausted, and burned through the detonator to fire the charge. Rockets were used at The Smalls, Heligoland and Tusker Rock in Ireland, although by the turn of the century this type of signal had been generally superseded. At Flamborough Head rockets were replaced by a siren in 1908.

The rockets provided Lundy with an all-round signal, reliable for two or three miles and they remained in use until 1897, when the new lighthouses were built. If the rockets had still been in use in 1906, Adair may have heard them, although the ten minute intervals between firing were too long to be of much use for fast ships. If the cannons had still been in use in 1906, Adair would surely have heard them as the *Montagu* grounded a mere three-quarters of a mile from the battery!

The two-note siren installed at Lundy North in 1897 remained in service for nearly thirty years before the station was refurbished. The fog signal building programme, undertaken by Matthews and Rayleigh, slowed down during the War years. By the time of Rayleigh's death in 1919, the Engineer in Chief was David Hood who succeeded Matthews to the post in 1915. Hood is remembered for the

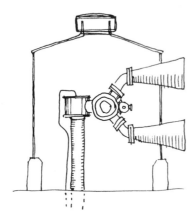

*Section through iron turret showing
twin resonators on the 12-inch siren.*

introduction of improvements to pressurised vapour illuminants, but he only held the post for a few years and died in 1924. His successor was John Bowen, who remained Engineer in Chief to Trinity House for the next thirty years.

After Rayleigh's experiments at Trevose Head in 1913, there had been little further progress with disc sirens but in 1923, an experimental siren was tested at Dungeness, just prior to Bowen taking over as Engineer in Chief. Under John Bowen, investigations into improved motor-driven sirens proceeded. The Dungeness device may have been a prototype of the instrument subsequently developed under Bowen: the twelve-inch diameter, motor-driven, cylindrical siren.

John Bowen was responsible for re-equipping many lighthouses, sometimes installing a new fog signal and usually replacing the old engines. The new twelve-inch siren incorporated one of Rayleigh's ideas: two trumpets were superimposed in a vertical alignment with the horn axes displaced by one half wavelength of the siren note, to augment the strength of the signal.

Cast-iron conical resonators were positioned one above the other in an iron turret placed on the top of the fog signal building. Start Point had the new turret and twelve-inch siren in 1928, when the five-inch sirens and upright bell mouthed trumpets were discarded. The following year, Lundy North was similarly re-equipped, as was St Catherine's Point in 1932. Bardsey Island followed soon after, but not until 1952, shortly before Bowen retired, did the Longstone lighthouse receive its twelve-inch siren, during the major rebuilding to repair the damage sustained during the War.

Pendeen lighthouse in Cornwall was the first to be equipped with a twelve-inch siren in 1925, when the station was refurbished.[7] The cylinder of this siren was

---

[7]   Pendeen Order Book shows that the new Gardner engines were in the 'process of being installed' in June 1924, although the change to the fog signal did not appear in the List of Lights until 1926. John Bowen and his assistant from Blackwall, Mr Swailes, visited the lighthouse in June 1926 to examine the completed refurbishment, which also included a new electric filament lamp.

The 12-inch siren used at Pendeen.

seven inches deep with sixteen circumferential ports. The ports were square cut, unlike the bevelled edges of the automatic sirens and measured 5 inches long by 1 inch wide. The siren was driven up to speed by a small air motor. Pendeen originally had twin five-inch sirens and the equipment and trumpets were the same as at the Lizard. When the new siren came to be installed in 1925, the original sounding receiver, siren pedestal and assembly were largely retained, along with the trumpets. The upper parts of the pedestal were modified to take the size and weight of the new instrument. The new siren chest had to fit in between the original trumpet throats. The twelve-inch siren at Pendeen was inverted relative to the orientation of the five-inch cylinders and the cast iron chest had two large ports on either side, leading to the modified trumpet throats. The full diameter of the siren was open to the air pipe from the admission valve. Above the siren itself, in the little remaining head space, was gearing to link the siren shaft to the drive shaft of the air motor. The siren was set spinning when the actuating machinery opened a valve, which started the air motor about five seconds before the main air was admitted to the siren. The siren sounded a seven-second blast every minute.

With the five-inch siren air was effectively forced through the cylinder walls and out of the open diameter of the cylinder into the trumpet throat. With the new siren at Pendeen, air was fed into the bottom of the siren through the open cylinder mouth and out through the ports in the walls of the rotor and stator into the trumpets. The siren design had equal widths between ports and bars and the idea behind this was to give a distinctive cut off to each burst of escaping air. Unlike the standardised five-inch, all of the twelve-inch sirens differed in the size and number of ports in the stator and the throat.

The siren made for Start Point was eight inches deep and had two sets of sixteen ports, 2¾ inches by 1 inches, but with four narrower ports over the throat area.

117

This siren was installed early in 1928, in the new turret attached to the roof of the original, circular fog signal building. New engines and extra air storage receivers were installed. Unlike Pendeen, the siren lay horizontally inside the turret. The trumpets – more prosaically 'conical resonators' – were made of cast iron and the throat of each attached to the horizontal siren chest, so that the mouth axes were separated by one half of the wavelength of the note, about three feet.

As previously noted, the layout of the engine room at Pendeen was very similar to that at Lundy North, but when the twelve-inch siren arrived at Lundy it was installed, not into the earlier siren pedestal as at Pendeen, but in a roof turret as at Start Point. Whilst work was underway in early 1929, at Lundy North, one of the old Hornsby engines was retained to sound a temporary signal. Installed outside at the front of the building was a single siren, with baby Rayleigh horns and two air receivers placed close by.

Increased air storage was created inside the building for the new siren by the addition of three more air receivers, comparable to the pair from 1897, which remained in their original position. Two new ones were placed hard up against the wall of the workshop and a new sounding receiver replaced the old siren equipment in the bay at the front of the building. The sounding receiver supplied air at 25 psi through a six-inch pipe directly to the siren in the turret above. The siren and resonator arrangement was as described at Start Point and the new character was a 4½ second blast, 2 seconds silent, a 1½ second blast, and 82 seconds silent.

The dimensions of the conical resonator were calculated using a formula which stated that the length should be half the wavelength of the note plus a quarter of the diameter of the mouth. It was still recommended that the siren should then be tuned to the fundamental frequency or else the octave, or perhaps even the twelfth of the note, whichever responded most satisfactorily.

Despite new developments in radio and electric signals, Bowen reckoned that compressed air signals would remain important for some time to come and since '… the best fog signal at its best is inefficient', it was important, therefore, to 'get the most you can out of an inefficient instrument …'.[8]

More powerful engines were required and were, by then, available. Gardners of Manchester offered nearly a hundred different types of engines in the early years of the 20th century and their factory at Patricroft produced over fifty engines every week at that time. In 1910, they first introduced a vertical engine and these semi-diesels were subsequently developed from 1913. These engines had a low compression ratio and as compression alone would not produce sufficient heat to fire the mixture, they were started with blow lamps like the earlier hot bulb engines, such as the Hornsby previously used at Lundy North. The Gardner VTs were two-stroke engines and once started, the blow lamp was removed and the 'hot bulb'

---

[8]   Report on the International Lighthouse Conference, 1929.

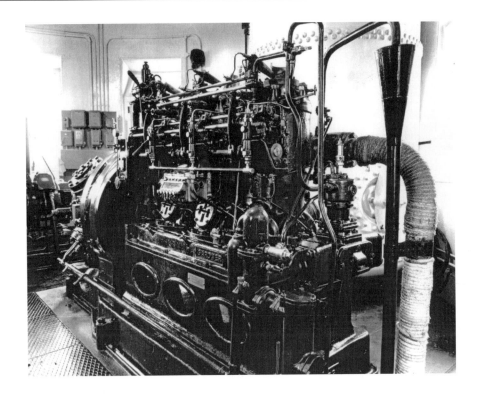

One of the Gardner T engines at Start Point, like those used at Lundy North. Courtesy Trinity House.

then maintained sufficient heat so that even with their low compression, it was sufficient to fire successive charges.

The T-type engine superseded the VT and was made with various sizes and numbers of cylinders. These engines were still quite slow running, between 290 and 450 rpm, but some could produce 300 bhp. Trinity House placed orders for several 72 bhp four cylinder engines, some of which were installed in lightvessels, but the two engines which were purchased for Lundy North in 1929, were three cylinder engines like the ones installed at Start Point in the previous year.

Each of the three cylinder domes had a burner directed on to it, attached to a pressurised paraffin supply. The starting burners were supplied with compressed air from the main receivers and were directed on to the cylinder domes for five minutes. Air to the starting burners was then turned off, but turned on to the air starting valve at the top of the cylinder and the compression cocks were closed. The cylinder was primed with fuel by pressing the charging levers and the governor control lever was set into the 'starting position'. The engine began running when

119

the starting lever linked to the cylinders was briefly raised by the keeper. To vary the running of the engine, the fuel could be directed into the coolest or hottest parts of the combustion chamber. As with the Hornsbys, during the time taken to start the engine, the siren was operated by stored air.

At Lundy North, the two Gardner T engines occupied the same space as the Hornsbys but the bed-plate was slightly longer as the Reavell Quadruplex compressor was directly driven by the engine via a friction drive wheel on the crankshaft. The Reavell Quadruplex compressor was first used at a fog station in 1904 and, essentially, the same type of machine is still used at Alderney, Nash and the Lizard.

In 1969,[9] the Gardner 3T engines joined the collection of engineering history residing in the watery archive off many lighthouses. It is thought that one of the engine house walls had to be dismantled to facilitate the departure of the huge Gardners. The value of the gun metal may have saved the siren from a watery fate, but probably only to be melted down later.[10]

New diesels were installed in 1969 to generate power for the 72-Tannoy-speaker electric fog signal which replaced the twelve-inch siren, but this is also now part of the history of the site. The lighthouse is automatic and has been solar powered since 1991. The curved 1969 fog signal bank did not much detract from the elegant lines of Matthews' 1897 building but the solar conversion is utilitarian rather than graceful.[11]

Lundy South lighthouse was manned until Christmas 1994. Shortly before automation, the supertyfon air signal was replaced by an ELU 800 electric emitter.

Attempts to salvage HMS *Montagu* in 1907 were unsuccessful, though some valuable items were recovered. Attempts to plate over the holes and pump the vessel with air only succeeded in floating the ship further into the cliffs. The remaining tons of ironmongery now rest in peace – Lundy North is silent and Lundy South only issues a high frequency peep.

[9] In April 1969, the modernisation of Lundy North began with the temporary installation of an electric fog signal. Trinity House Notice to Mariners.

[10] The Trinity House National Lighthouse Centre in Penzance has twelve-inch sirens from Start Point and Longstone.

[11] A small plastic optic has been installed on top of the disused tannoy stack, with solar panels propped up behind it. This simple arrangement provides a satisfactory light and saves on generators, diesel and engineers' visits.

Chapter 9

St Abb's Head
The Scottish siren and the radio beacon

The 1876 'Brown' siren appears to have remained in operation at St Abb's Head for over thirty years and if all of the original equipment was still working in 1910, when the 32-inch Ericsson hot air engines were replaced by oil engines, the initial outlay was justified.

The St Abb's siren differed from the one used in the South Foreland trials in several respects. It used compressed air, not steam, and the trumpet was upright, resembling the reed horn design rather than the long horizontal horn used at South Foreland. In addition, the St Abb's siren rotated at a slower speed than the South Foreland instrument so that it produced a lower note of 244 Hz. The range was impressive, as keepers at the Isle of May over twenty miles away often heard the siren. St Abb's sounded a six-second blast every seventy-five seconds. The siren was mechanically driven by the hot air engine and as the sounding air was not admitted until the disc was spinning, the sound produced was a sustained, even note.[1]

In America, the early motor-driven siren did not easily maintain an even running speed, sometimes resulting in notes of varying pitch, though there are no known complaints relating to the St Abb's siren. The automatic siren was less likely to produce these effects. The American instrument at St Abb's was the first siren on the Scottish coast, but by 1880 there were four others in service. By 1900, there were twenty-three sirens and by 1930 there were sixty, including those operated by port authorities. Fifteen new lighthouses were built in Scotland after 1900, nine of them with sirens.

After the first automatic two-note siren was installed at the Mull of Kintyre in 1883, a further seventeen Scottish lighthouses had two-note sirens, although the last four, all installed after 1900, were all of the new seven-inch motor-driven type. These were at Girdle Ness, Killantringan, Bressay and the Butt of Lewis.

---

[1] The 12-inch diameter disc was designed to spin at 1464 rpm.

The Girdle Ness siren, included in the 1901 fog signal trials at St Catherine's, had 6 ports which produced the low note of 100 Hz; the high note of 234 Hz was produced by 14 smaller ports.[2] The fog signal trials could not establish the superiority of the Girdle Ness siren over the others tested and although motor-driven sirens performed well in the trials, the merits of a sound produced by an automatic or motor-driven siren could not be absolutely determined.

Engineers working for Trinity House and the Northern Lighthouses had faith in their own designs and took a pride in their work which was tinged with a national competitiveness! Since the design, installation and operation of fog signals was not an exact science, the implementation of particular fog signal designs owed something to individual professional judgement and the nature of the decision-making processes within Trinity House and the Commissioners of Northern Lighthouses.

In general, after the 1901 trials, the automatic five-inch siren was used in England and Wales and new sirens installed in Scotland were of the seven-inch motor-driven type. In Scotland, the automatic and motor-driven two-note sirens were replaced by the single-note instrument from 1914 onwards, although the ongoing refurbishment programme continued through both World Wars, so that one or two of the high- and low-note sirens survived into the 1950s.[3] A seven-inch single-note siren was installed at St Abb's Head in a new horn house in 1910.

David A. Stevenson was the Engineer to the Northern Lighthouses from 1887 to 1938 and a drawing signed by him, dated March 1910,[4] shows the fog signal equipment intended for St Abb's. Two standards support a heavy casting with two admission valves and a central chest for the siren. The equipment in the drawing closely resembles that installed at the Mull of Kintyre in 1898, but the Mull siren was a two-note instrument with the chest divided for the high and low note ports. The St Abb's drawing suggests that this chest was not divided, and that only one admission valve was designed to be operational.

The equipment still on site at St Abb's differs from the above, indicating a change to the initial proposal or a later replacement of the original equipment.

Steven & Struthers, a Glasgow engineering company, produced boilers, valves and other related equipment, primarily for the shipping industry. From the 1880s their brass foundries manufactured ships' whistles and sirens and in the 1890s, the

---

[2]   Both notes sounded by the Girdle Ness siren were lower than the earlier Holmes type two-note sirens which sounded at 280 Hz and 640 Hz.

[3]   All of the lighthouse authorities spread capital costs and maintenance budgets to provide a rolling refurbishment programme. Different equipment was used to suit the individual conditions prevailing at various sites so that for many reasons, some items of machinery outlasted others. Refurbishment could involve the removal of virtually all of the previous plant or might sometimes entail just a few replacements. The usual lifespan of fog signal equipment was about 25 to 30 years, but some engines lasted for 40 or 50 years – the Hornsby engines at Nash Point were installed in 1904 and replaced in 1968! Original fog signal equipment lasted up to 100 years in some cases.

[4]   NLB drawing C/68/37. The National Monuments Record of Scotland holds many NLB engineering drawings.

St Abbs pedestal and siren chest minus the siren.

company was awarded contracts to equip new lighthouses on the Scottish coast.[5] The company undertook work on the fog signal at the Mull of Kintyre in 1898 and other contracts included Tod Head, where they installed 9 hp Campbell gas engines and Stroma, where they installed a two-note siren to the specifications of David A. Stevenson.

During the 1930s, Steven & Struthers were awarded contracts for refurbishing work at Tod Head, Noss Head, Fair Isle North and St Abb's Head. Stevenson's 1934 work specification[6] for St Abb's Head required the contractor to complete all work within five months and stipulated that the fog signal character was to be altered on the changeover day specified in the Notice to Mariners. The work in hand was not to interfere with sounding the siren and one of the new Atlantic engine and compressor sets was to operate the fog signal, whilst the old oil engines were removed and other work in the engine room was undertaken.

---

[5]   Stevens & Struthers installed sirens and other equipment at Stroma, Noss Head, Tod Head and Mull of Kintyre. Between 1896 and 1898, sirens were installed by other companies at Rattray Head, Pladda, Corsewall, Mull of Galloway, Cloch and Isle of May.

[6]   Unpublished specification of work for St Abb's Head Fog Signal drawn up by D.A. Stevenson, 14th July 1934.

David A. Stevenson and his father, also David, and uncle Thomas, tried a variety of different oil and gas engines over the years including Priestmans, Crossleys, Campbells amongst others. From the 1920s, Scottish fog signal installations were usually equipped, or re-equipped, with Atlantic engines.[7]

The four-cylinder Atlantic engines had a piston diameter of 5½ inches and a 7-inch stroke. They were designed to run on lighthouse paraffin, although they were started on petrol. Each cylinder had a spark plug connected to the magneto and the engine ran for five to ten minutes on petrol, before the keeper switched a lever to shut off the petrol to the carburettor and isolate the high tension leads. The hot cylinders then enabled the engine to continue running on paraffin. The engines ran at 450 rpm.

Three sets of Atlantic 4 DP engines were installed at St Abb's, with Alley & MacLellan compressors directly coupled to the engine on the same bed-plate and driven by a centrifugal clutch on the flywheel. The compressors had two large cylinders with reciprocating pistons and delivered up to 150 cubic feet of air per minute at 25 psi.

In service, two of the three engine and compressor sets were used, with the spare set alternated for even wear. Records were kept of the engine set used each time and the total duration of the fog signal operation was recorded automatically on a circular chart. James Ritchie of Edinburgh, who also produced fog signal clocks, supplied the fog signal recorder which was operated by the compressed air in the system. The charts were calibrated in 24 hours and were retained with other station records.

Stevenson's work specification for St Abb's Head also included improvements to the water tank, which circulated cooling water to the engines and compressors and the installation of new pipework for water, air and fuel. This pipework was largely suspended from roof hangers in the engine room and in addition to the engine exhaust pipes, there were small bore, copper pipes connecting the 100-gallon paraffin tank and 5-gallon copper petrol tank with the engines. Also, there were one-inch supply and two-inch return pipes linking the engines and compressors to the cooling tank and four-inch and six-inch iron pipes linking the compressors to the air storage receivers. An earlier air receiver was removed from the engine room and installed outside.

Air storage receivers were placed outside the engine rooms at Scottish lighthouses and the air was usually stored at the working pressure of 25 psi – fluctuation in the pressure was compensated for by the large number of receivers and the length of the pipework itself. Most Scottish sirens were installed some distance from the engine room. Between five and seven receivers was common practice; after refurbishment in 1934, Tod Head had nine.

---

[7] The Atlantic Engine Company Limited had their works at Wishaw.

The horn house on the cliff below the lighthouse at St Abbs.

The practice of storing high pressure air and supplying the siren via a reducing valve was less common in Scotland than at Trinity House fog stations,[8] not least because the petrol started Atlantic engines could quickly begin compressing air.

The three engine and compressor sets cost £1136 16s 9d but the total cost was £2179 12s 3d which included transportation, erection, six months' supervision and fitters' and labourers' pay, at 3s 6d and 3s 0d an hour respectively. Although the character of the signal was changed, work undertaken in the horn house did not include a new siren.

The lighthouse at St Abb's was first lit in 1862 and the lantern is 255 feet above sea level. The tower was built lower down the clifftops than the accommodation and is reached by a flight of steps. These buildings were joined, on the steep and precipitous clifftops, by the new 'Fog Signal House' in 1876 and additional keepers' quarters were built nearby within the next few years.

The current horn house stands about 75 yards from the main buildings, below the tower and right on the cliff edge. The cliffs are nearly 200 feet high at this point. Walking down to the horn house leads one past the cooling water tank and the site of the five air receivers on the left. To the right of the lighthouse tower there used to be two additional receivers. The six-inch air main still leads down the remaining thirty yards to the horn house.

---

[8]   Scottish receivers were tested to 40 psi. Where high pressure air was stored at Trinity   House lighthouses, pressures exceeded 150 psi.

The horn house, built in 1910, is typical of the design adopted from 1900. At lighthouses situated at low altitudes the design was sometimes elevated on a tower.[9] A large horizontal horn emerges from the centre of the sloping roof, the mouth of which protrudes over a curved wall which protects the front of the horn house. The radius of the wall is 8 feet and the horn house itself is 8 feet by 7 feet. The roof slopes from 5 feet 10 inches at the front of the building up to 6 feet 10 inches at the rear. The curved wall on the exposed cliff edge supports an iron track, on which the horn rests on its carriage. This carriage could be moved by a hand-cranked pinion to face the horn into the wind,[10] but this practice ceased many years ago, as the thickness of red lead oxide paint on the carriage and pinion testifies. The conical trumpet is 4 feet in diameter at the mouth and tapers to 8 inches in diameter at the throat and it too is thick with red paint.

> The trumpet design was largely unchanged from 1900. The diameter of the mouth is approximately half the wavelength of the siren note and the total length of the horn along the axis is 14 feet. The horizontal section is 10 feet long and the horn has a 3-inch flare around the rim of the mouth. A ball and socket mounting at the throat enabled the horn to be raised or dipped to make allowances for 'rising sounds', but the practice of dipping the horn lapsed with the introduction of facing the horn into the wind, as did covering the mouth of the horn to prevent rain and spray entering. Some work specifications from the 1930s still included a leather cover with a steel frame for this purpose, however. The cover blew open with the first siren blast and was repositioned by a rope and pulleys when the siren was stopped.

The sloping roof makes the interior of the horn house appear even smaller than it is. Below the entry point of the horn is the siren chest and admission valve. The siren has been removed from the chest at St Abb's, along with the actuating clock which used to be on the right-hand wall. The six-inch air main emerges from the floor at the front of the horn house and connects to the admission valve.

The 1910 siren retained the old character of 6 seconds blast, 69 seconds silent, but the character was changed at the subsequent refurbishment to give one blast of 3½ seconds every 45 seconds. The actuating clock opened the main air valve to sound the siren about 10 to 15 seconds after starting the air motor spinning. The clock had an aluminium air turbine motor which automatically re-wound the falling weight. The clock sounded a quick and resolute 'tock, tock, tock'.

---

[9] Towers were built at several sites where the low elevation above sea level was thought to be detrimental to sound transmission. Towers were constructed for the new fog signals at Tod Head and Stroma lighthouses in 1934 and the latter tower remains beside the original 1897 hexagonal horn house. The hexagonal design horn houses were built for the new sirens installed in the 1890s, although the horn house at the Mull of Kintyre, the earliest surviving example, dates from 1883.

[10] Moving the horn to face into the wind was recommended by Tyndall after the South Foreland trials in 1874. Early Trinity House sirens were similarly designed, see Chapter 4.

The horn supported on its carriage at St Abbs (left), and right, the clockwork actuating machinery at Corsewall lighthouse.

Within an elbow joint in the 1½-inch pipework from the clock, a rotating spindle opened air to the siren motor and in another elbow joint was a spring-loaded valve, which was depressed by a cam wheel geared to make one revolution, according to the character of the signal. The heavy 'tocking' continued throughout the hissing noise, which started as air was directed through a 5/32nd inch nozzle on to the vanes, or 'buckets', of the air motor on the bottom of the siren.[11]

The jet of air started the heavy siren spinning on its bearings and as the siren was driven up to 1200 rpm, the spinning vanes interrupted the air jet to produce an audible rising note inside the horn house. Once the correct speed was attained, the centrifugal governors in the base of the siren retarded any further increase and after a few seconds the clock tripped the actuating valve to open the main admission valve and release air through the siren for the 3½-second duration blast.[12]

It would be loud enough inside the horn house, but the intensity of the sound emanating from the mouth of the horn at nearly 130 decibels was not only painful on the ears but nauseously experienced by most of the body's internal organs. Fortunately, the automatic operation of the equipment in the horn house did not normally require exposing the keepers to this ordeal.

With a single blast at St Abb's, the fall off in pressure throughout the system was slight, but at stations with long and multiple blasts, the first blast – at 25 psi – could be up to 7 psi greater than the final blast. The siren at St Abb's used about 350

---

[11]   The small conical buckets were drilled at such an angle that the jet of air blowing on to the rim of the cylinder caused the siren to spin. Designs varied, for instance, on the two Isle of May sirens, one had 48 'buckets' and the other had a 32-bladed turbine. The blades or vanes were more common than the buckets.

[12]   The air motor ran throughout the blast and for an additional ten seconds before and afterwards. At some stations, the characteristic required the motor to cut in to keep the siren free wheeling between soundings.

The siren in the chest showing the vanes used to set it spinning.

cubic feet of air with each blast.[13] Although some way from the engine room, the sound of the compressors would be audible in the horn house as their labour was conducted through the pressurised pipework: the pipes and receivers would hum a chord to the 'chugging' of the two compressors!

The seven-inch diameter siren was made of gunmetal and had an overall length of 20 inches from the top rim of the cylinder to the ballbearings at the bottom of the shaft. The cylinder itself was 7½ inches high. The siren chest was made of cast iron and had an internal, annular space surrounding the siren ports which was nearly 12 inches in diameter. The siren was lifted into the chest through an 8-inch diameter aperture in the bottom of the casting and securely bolted into place. When the admission valve was opened, air filled the annular chamber and flowed through the ports in the siren.

The chest and the adjacent admission valve was supported by two cast iron standards at a height of 3 feet above the horn house floor. Attached to the left side standard was a counterbalanced 'lifter' used to remove and replace the siren in the chest. Keepers were required to periodically examine the siren and replace it with a spare one for routine maintenance.

---

[13] The powerful Scottish sirens used large quantities of air. The sirens had large port areas and often gave long blasts and in some cases, the character of the signal required multiple blasts. Duncansby Head, for example, gave a blast of 2½ seconds followed by silence for 2½ seconds; blast 2½ seconds, silence 2½ seconds; blast 2½ seconds; silence 2½ seconds; blast 2½ seconds; silence 2½ seconds; blast 2½ seconds; silence 97½ seconds. As with all air signals, the character was limited by the ability of the compressors to replenish the receivers within the interval between soundings or else power would diminish, ultimately preventing the siren from operating.

The Scottish sirens had different manufacturers[14] but the design remained fairly consistent over the years. The instrument varied according to the required character, but the siren typically had six ports about an inch wide and five inches long. The size and number of ports could be supplemented by variations in the position and number of ports in the rotor and stator, but usually the same pattern was cut in both.

The Girdle Ness siren, tested at St Catherine's in 1901, sounded high and low notes; the low ports were 2 inches long and 0.81254 of an inch wide. As the 7-inch siren was developed and high-note ports were dispensed with, the length of the six low-note ports was extended in later designs. The Girdle Ness siren ports were cut with bevelled edges which increased the speed of rotation above the motor speed, once the sounding air was admitted. This would have produced a slightly rising note although much less pronounced than the 'whoop' of the Trinity House, five-inch automatic siren. It is possible that other Scottish sirens produced a momentary slight rise in pitch at the start of the blast. The top edge of the siren ports was also cut with bevelled edges to reduce undue stresses when the air was admitted and the siren was lifted in its bearings by the force.

The work specification for St Abb's was drawn up in July 1934, but the work was postponed and only completed in 1938. In March 1938, David A. Stevenson retired from the post of Engineer to the Board and died shortly afterwards. Although his death ended the 140-year association between the Stevensons and the Northern Lighthouses, his brother Charles and nephew D. Alan worked for the Clyde Lighthouse Trust and maintained the family connection with lighthouse engineering.

D. Alan was responsible for a unique development of the radio beacon, although the first radio beacon in Scotland was installed by his uncle, David, in 1929. David Stevenson had plans to install a 'wireless beacon' at St Abb's in 1933.

Marconi carried out many of his early radio transmission experiments from or near lighthouses, including St Catherine's, South Foreland and Flatholm and by 1910, early experiments were under way with the idea of a radio beacon – a navigation aid capable of providing a position fix even in the thickest fog. These experiments were initially made using wavelengths of 1000 metres.

The development of radio for navigational purposes accelerated during the war years, with the result that by the early 1920s, radio beacons were installed on the American coast and were shortly followed by similar systems in Europe.

In Britain, the Marconi Wireless Telegraph Company developed the equipment necessary for generating and transmitting a signal, along with that necessary for receiving it aboard ship, but other companies were similarly involved, including the German ones, Telefunken and Pintsch. Pintsch already had a history of lighthouse engineering, particularly with gas lighted buoys.

---

[14] The sirens were made by Alexander Westwood of Glasgow, Stewarts of Glasgow, Steven & Struthers of Glasgow and Milnes of Edinburgh. These companies won different installation and refurbishment contracts over the years.

At the International Lighthouse Conference held in London in 1929, it was recognised that radio beacons were going to be of major significance to navigation and that the development of radio beacons was sufficiently advanced to require a specific conference which would address the issues of parity and performance. The aims of this conference, held in 1931, were to agree on operating systems and evaluate potential future developments.

The manufacturing companies involved were keen to protect their own interests and investments, although it was necessary to establish certain agreements if the developing systems were to be of any use. The separate development of the new technology in different countries had resulted in different transmission frequencies being used. The German beacons transmitted between 300 and 600 kilohertz (300,000 to 600,000 Hz), whilst in Britain the chosen bandwidth was between 800 and 1200 kHz. The French decided on a wide band, between 600 and 1800 kHz.

It was not only transmission frequencies which needed to be agreed upon, but also the appropriate note of a transmitted radio signal: it was important that the radio operator could easily identify the sound above the hums and buzzes inherent in the system. For a time, a similar debate ensued to the earlier one on the relative merits of the high and low notes produced by the siren.

Some decisions could not be made by the delegates to the radio beacon conference as this responsibility was not in their hands, but subject to the authority of the International Consultative Committee for Radio Communications. The United States, for example, was determined to use a specific frequency range, between 285 and 315 kHz, but this bandwidth was increasingly being used for broadcasting by radio stations in the late 1920s and early 1930s. By 1930, there were sixty radio beacons in the United States and new development required international agreement on bandwidths, transmission times and the nature of the transmission.

In principle, those beacons in operation were performing two functions, firstly as a 'wireless fog signal' and secondly, as a 'directional wireless' enabling the mariner to obtain a position fix. The frequency of a specific beacon could be tuned into aboard ship and a direction finding aerial was then moved to detect the strongest and weakest signal from the beacon. The weakest signal occurred at ninety degrees to the bearing of the beacon. With a little experience, the bearing of the beacon could be calculated and references from one or more other beacons could give an accurate position fix when plotted on a chart, by day or night or in fog.

In practice, the various systems in use had been developed with slightly different ambitions. Some system designers considered the information content and the nature of the transmission; others, the simplicity of operation. Others were motivated by the overall cost of the receiving equipment; if too expensive, shipowners might be deterred from investing in it.

The Marconi system used a rotating framework to transmit short wave Morse code signals and the RAF used similar equipment for position fixing. Several countries followed this route, but the inconvenience, or even impracticality, of the cumbersome transmitting and receiving equipment required was underlined by the simplicity of the system adopted in America and favoured by the British lighthouse authorities. This system was used by the Americans since the early 1920s and established itself as the standard, even though it required specific equipment to be installed on board ships. The system was also relatively cheap and efficient for the lighthouse authorities to operate!

The American beacon system did not require specialist radio operators and the beacon simply repeated a single Morse code letter assigned to identify the particular station. This simplicity was lacking from European beacon transmissions which tended toward overloading the channel, furnishing the specialist radio operator with more information about the direction and distance of the transmitting station, but at the cost of more elaborate equipment being required on board ship.

As with lighthouses and fog signals, distinguishing one radio beacon from another was very important, but it was additionally necessary to prevent interference between signals. It was for this reason that the narrow bandwidth for particular beacons within a certain area was specifically agreed. It was also necessary to co-ordinate the time intervals between separate transmissions from different stations and Trinity House adopted a policy of transmitting for one minute in every four during fog, reducing the transmissions to three every half hour in fine weather. The equipment used produced a sound signal between 800 and 1200 kHz with a wavelength of between 950 and 1050 metres. A comparable system was developed in Germany by Telefunken.

At the lighthouses, 50-volt batteries were used to power the radio valves in the signal generating equipment. The masts themselves, experimentally constructed in steel, aluminium and other less corrosive metals, sprouted at various sites and were up to 100 feet high.

The first radio beacon in regular service in Scotland was constructed at Kinnaird Head early in 1929 and was shortly followed by another at Sule Skerry. 'Wireless beacons' at North Ronaldsay and St Abb's Head were planned in 1933. Separate and distinct from the radio beacon systems, radio communication equipment was also introduced to enable contact with distant stations. The Butt of Lewis lighthouse had Marconi radio equipment from the mid 1920s.[15]

Radio was also used on the Clyde by David A. Stevenson's brother Charles, who introduced radio-controlled fog signals – acetylene fog guns were remotely switched on and off as required. Charles Stevenson and his son, D. Alan, as engineers to the Clyde Lighthouses Trust[16] were responsible for a distinctive development of the

---

[15] 'Radio telephones' were introduced by Trinity House in the 1920s at several lightvessels, to replace the earlier submarine cable connections with the mainland.

[16] The Clyde Lighthouses Trust was incorporated into the NLB in 1956, but the Clyde Port Authority regained responsibility for navigational aids in the River Clyde and its approaches in 1966.

radio beacon in which they synchronized the radio transmission with a fog signal blast.[17]

The development of the radio beacon convinced many people that the days of the siren, and possibly the lighthouse itself, were numbered, but D. Alan Stevenson recognised that sound signals would continue to be of importance for some time into the future. He also believed that the necessary installation of expensive equipment aboard shipping might detract from the advantages of the radio beacon system and he proposed a practical and economical alternative development in the prototype 'Talking Beacon'.

By 1929 a synchronized radio and diaphone signal was being tested at Cumbrae lighthouse and the system was fully operational from 1931. It was similar to the submarine signal system in principle[18] in using the time difference between synchronized signals to indicate the distance from the source. As with other radio beacons, the mariner could obtain a position fix on Cumbrae but in addition, the conventional fog signal helped to provide an easily understandable indication of the distance of the vessel from the signal. Taking the speed of sound in air to be about 1120 feet per second and the speed of radio waves to be instantaneous, if the radio and air signals were synchronized at transmission, the delay between the reception of both aboard ship related to the distance travelled by the sound. Therefore, the distance of the vessel from the station could be calculated. In the trials, the mariner had to time the interval with a stopwatch, as was often done when proceeding in fog, when the echo of the ship's own siren from the cliff or other obstacle was timed. A five-second delay indicated approximately one mile from the station, ten seconds was two miles and so on.

The Talking Beacon at Cumbrae used an automatic voice recording which announced the name of the station and advised which blast of the diaphone was to be used as the reference. The voice then rapidly counted 'one, two, three, four, five, six, seven, eight, nine … [ting] one mile; one, two, three …' and so on. A metallic 'ting' coincided with the voice stating the 'mile' every five seconds. If the mariner heard the blast of the diaphone at the point where the voice over the radio said 'five', the vessel was approximately five cables, or half a mile, from the lighthouse.

---

[17] At Cumbrae lighthouse, a powerful diaphone fog signal was installed in 1924, one of the earliest in use in Britain and believed to have been made by Steven & Struthers under licence from the Canadian patent owner. Steven and Struthers refurbished the engine house and installed nine horizontal air receivers which stored air at about 40 psi. The Campbell oil engines dating from 1900 remained in service for compressing air for the new fog signal and for driving a dynamo to generate electricity for the radio transmitter.

[18] See Chapter 10 which deals with submarine bells. The idea of synchronised signals must have occurred to Tyndall when, at South Foreland, he saw the puff of smoke from the guns some time before hearing the report. He would have calculated the distance off the Foreland by comparing the time lag with the speed of sound.

A similar system was used at the Cloch lighthouse, where the distance off was announced in cables.[19]

The Stevenson invention was used at stations in America[20] but was not developed elsewhere in Britain or Europe. Criticism of the system centred on the essential problem of sound signals in air, namely, the reliability of transmission could be unpredictably interfered with by the wind and the phenomenon of 'zones of silence'. Other criticisms related to potential confusion of the synchronized start of the radio and diaphone signal. Sound-in-air transmission problems might render one or more of the blasts inaudible, so that incorrect correspondence between diaphone and radio signals might lead to positional errors.

Stevenson acknowledged the stated difficulties but maintained his belief that the simplicity and effectiveness of the Talking Beacon outweighed its disadvantages. Even with a strong wind across the signal, he thought the distance finding accuracy of the Talking Beacon was still better than that plotted from a conventional beacon, where minutes could be lost in calculating a ship's position on a chart from the information received. Without the need of a stopwatch, Morse code or calculation, the mariner was quickly and simply provided with information about his distance from the named Talking Beacon fog station. If the diaphone could not be heard at all, the beacon could still provide a position fix since an elemental portable direction finding set costing about £20 was adequate. The standard direction finding equipment designed for operation with the radio beacon system cost upwards of £200. All passenger ships over 5000 tons were legally obliged to be fitted with radio beacon receiving equipment from 1931, but for other vessels this was optional.

Policy and practice encouraged the development of radio beacons, and in Britain the Talking Beacons remained unique to the Clyde, although other kinds of synchronized signals were tried elsewhere.[21]

During the war years, lighthouses were discretely used according to secret notice given of scheduled convoys and radio beacons were discontinued. Many lighthouses suffered war damage and the beacon service was not immediately resumed after the war.

In 1951, a Geneva Agreement established new practices relating to call signs and transmission times for radio beacons, although many stations had already adopted the agreed transmission cycle, which was one transmission in every six minutes. The agreed signal consisted of the Morse code call sign repeated four times, followed by a long dash of 25 seconds. A further two call signs were followed by a 5-second period of silence totalling 60 seconds. The transmissions were made automatically. An alternator was started by electric motors before each transmission

---

[19] 'Cloch Point lighthouse speaking … when you hear the start of the second blast of the fog signal the distance from the lighthouse is stated on the radio in cables …'.

[20] Over twenty-five talking beacons were in use in the USA before World War Two.

[21] See Chapter 10 regarding synchronised siren and submarine bells, oscillators and radio beacons.

*One of the Kelvin engine and Alley Maclellan
compressor sets at the Mull of Galloway.*

time. When the coder machine started a motor in the sender unit, notches cut in the circumference of a disc opened and closed contacts as the disc revolved to produce the Morse character for the transmitted signal.

The 1951 Agreement also required some alteration to operating frequencies and a further change was to establish continuous transmissions. Many of the changes appeared in Notices to Mariners issued in 1953, including those relating to two letter identification signals. St Abb's radio beacon retained a three letter call sign, but it was changed to SAB and transmitted on 284 kHz. However, later reorganisation of the radio beacon groups no longer required the beacon at St Abb's and it was discontinued.

In the engine room at St Abb's, the Atlantic engines were replaced by Kelvin K2 diesel engines in December 1955. Like the Atlantics, these engines were also started on petrol and they remained in use with the fog signal until August 1987, when the siren was permanently discontinued,[22] along with many other Scottish sirens.[23] No Atlantics or Sentinel compressors survived in service, but Kelvin's and Alley MacLellan compressors were in use at nearly all of the Northern Lighthouse Board sirens until the fog signals were discontinued. St Abb's was used for experimental trials with fog detectors in the early 1960s, along with a radar beacon (RACON), which was permanently established at the lighthouse in 1968.

Many lighthouses and lightvessels once provided information to the Meteorological Office and St Abb's was one of the coastal stations included in the shipping forecast bulletins. I realised that the automation programme had reached another milestone when the announcer regularly omitted 'St Abb's Head' from the 'Weather Reports from Coastal Stations'. The lighthouse became fully automatic in 1993.

---

[22] From noon on 14th August 1987, the following fog signals were discontinued: St Abb's Head, Duncansby Head, Holburn Head, Holy Island, Corsewall Lighthouse. From noon on 30th November the following were discontinued: Girdle Ness, Kinnaird Head, Dunnet Head, Strathy Point, Cantick Head, Sumburgh Head, Eilean Glas, Killantringan, Mull of Galloway, Langness, Maughold Head.

[23] No Scottish sirens remain in service but some equipment is still on site at Kinnaird Head which now forms part of the collection of Scotland's Lighthouse Museum.

Chapter 10

The Seven Stones Lightvessels

A survey of lightvessel fog signals

The Trinity House steam vessel *Mermaid* worked from the depot in Penzance in the early years of the 20th century, carrying out reliefs at the rock lights and delivering supplies to lighthouses along the coast as far east as Start Point and Berry Head.

The area had over 50 navigation buoys in the early 1900s, including those in river mouths and port approaches, although only eight were lit by gas. Four had bells and one or two had whistles. The black and white striped bellbuoy at the Manacles rocks had a gas light and numbered amongst those serviced by the *Mermaid*. Buoys were repaired and painted in the depot on the quayside at Penzance. The depot was built on the site obtained during construction of Wolf Rock and the buoy workshop now forms the Trinity House National Lighthouse Centre.

There were still many old sailing vessels in use at the time, some much older than the famous Cutty Sark, (itself over 30 years old), which could be seen off the Lizard and around Falmouth into the late 1920s. A number of these vessels came to grief off the Cornish coast and one of *Mermaid*'s other duties was to search out and survey wrecks, either to mark them with buoys or disperse them with explosives if they were a hazard to navigation.

The *Mermaid* delivered small oak casks of fresh water to the rock lighthouses along with drums of paraffin and sacks of coal.[1] Other deliveries included gas for Berry Head lighthouse, cases of charges and detonators for the explosive fog signals at Eddystone, Longships and Bishop Rock and sundry supplies of all kinds, including boxes of books and even furniture and uniforms. Empty casks and drums were subsequently collected and the boxes of books occasionally moved from one station to another.

---

[1] *Mermaid*'s successors, *Satellite* and *Stella*, still delivered water in casks in the 1940s and even into the 1960s, but in later years, water and diesel were supplied to lighthouses and lightvessels by pipes from the tender.

The *Mermaid* also serviced the Seven Stones lightvessel although, as at the rock lights, poor weather could delay these visits and postpone the crew reliefs.

The first lightvessel at the Seven Stones reef was placed in 1841, to the east of the rocks about 10 miles north-west of the Scillies and 13 miles west of the Longships lighthouse off Land's End. The reef stretches for about a mile in otherwise deep water, fully exposed to the large Atlantic waves. Some rocks are uncovered at low tide but the construction of a lighthouse on the site was ruled out as being too costly and difficult. Compared to other exposed sites upon which lighthouses were built, the extreme nature of this site can be easily appreciated.

Fully equipped, the first lightvessel placed at Seven Stones cost a total of £4416 8s. 7d. Unfortunately, the 80-foot long boat was swept off its moorings soon after being placed, although an improved anchor and longer mooring chain securely held the vessel in place after it was repositioned. A forty hundredweight mushroom anchor, with over three hundred fathoms of one-and-a-half-inch mooring chain, held the ship in position in forty fathoms.[2]

Purpose-built lightvessels were introduced from the 1830s and when Seven Stones was placed in 1841, there were twenty-five others in service around the English and Welsh coast. Most of them were in the Thames approaches and on the east coast, with two on the Humber, two on the Mersey approaches, two off the Isle of Wight and one in the Bristol Channel.

The number of lightvessels increased from five in 1800[3] to eleven in 1820. There were only twelve in 1830 but this figure had doubled by 1840 and by 1870, had nearly done so again, although not all were operated by Trinity House. Few other light vessels were in such deep water and none in such an exposed site as at the Seven Stones.

The 1841 ship had two top mast balls and fixed lights and the hull was painted red. A crew of six was permanently aboard, though earlier vessels in the Thames approaches had a mere four man crew. At Seven Stones, the master and mate alternately took one month on and one month off and the remaining crew were aboard for two months with one month off. Most of the crew lived on the Isles of Scilly.

The carpenter and lamp trimmers also had to be fully competent seamen as lightvessels could be driven off their moorings in strong winds and seas. The early lightvessels stowed sails which were used when the ship was moved off station by accident, or when replaced by another vessel.

---

[2]   Chains became available and reliable after about 1820, prior to which time ships and lightvessels used hemp rope for anchoring.

[3]   Robert Hamblin is credited with the idea of fixing a lightvessel in position and indicating the position on navigation charts: a specific configuration of lights would enable one vessel to be distinguished from another. The idea probably originated with David Avery. The first five lightvessels marked shoals and sand banks: The Nore, 1732 (Thames), Dudgeon, 1733, (The Wash), Owers, 1788, (Spithead Approaches), Newarp, 1791 (Norfolk Coast), North Sand Head (North Goodwin), 1793 (Kent/Thames Approaches). The original vessels were modified barges.

The original lightvessel had Argand lamps with silvered reflectors and the entire lantern was hoisted to the masthead by rope each evening and lowered for cleaning and preparation each day.

Robert Stevenson's lantern was designed to surround the mast and was raised and lowered by means of pulleys. This lantern became standard from the early 1800s. To assist identification, lightvessels had fixed lights configured in distinctive ways. A number of vessels had two lights and North Sand Head (North Goodwin), Newarp and Crosby had three, prior to the introduction of the revolving lantern mechanism. Until 1879, the Seven Stones showed twin fixed catoptric lights on two masts, at 20 feet and 38 feet.

From about 1840, lightvessels, including the Seven Stones, were equipped with a gong and a cannon[4] which could signal warnings in foggy conditions and alert a vessel spotted heedlessly heading into danger. The gun was fired as a last resort to attract attention, when the ball at the masthead was lowered to half mast until the endangered vessel changed course. However, the presence of a lightvessel was insufficient warning by itself for some captains, and as there were no agreed international conventions, shipping did not always recognize the warning signals.

The warning gongs were two feet in diameter, about four inches deep and weighed about thirty-five pounds. The gong was not struck like the bell that was also used on some vessels, but was steadily beaten to produce a sustained crashing sound rich in harmonics and with a fundamental note about C or D. The note varied with the size of the gong. There was no prescribed characteristic nor any kind of machinery involved in striking the gongs, and the job must have been very tedious. The Chinese gongs were imported from Hong Kong at a cost of £4 each (1870 prices!).

The Admiralty List of Lights records an early fog signal entry for the Owers Lightvessel in 1832: 'During fogs a gong is sounded every ten minutes'. Above the noise of the wind and sea, the gong would have been distinctive but its range was not great, possibly less than a quarter of a mile. However, the Warner lightvessel gong could be heard at the Nab by all accounts, three miles distant!

After thirty years' service, the original vessel was in need of serious attention and it was replaced by a new vessel in November 1871.[5]

One of the earliest compressed air fog signals used aboard lightvessels[6] was installed in the new ship and the reed horn superseded the gong as the standard fog signal at this station.

---

[4] Small carronade were also used aboard lightvessels. The term carronade is derived from the Carron Ironworks at Falkirk, which first produced this type of short range, large calibre gun from 1778.

[5] 'The new lightship for the Seven Stones, towed from Blackwall to her destination in 42 hours by the Trinity steamer *Galatea* is securely moored in her sea tossed berth'. *Cornish Times*, November 15th, 1871.

[6] The original Daboll reed at Newarp lightvessel dated from 1868. Other reeds were placed aboard Outer Dowsing and Spurn lightvessels in 1875 and 1877.

The Holmes reed and trumpet was operated by 1½ hp Ericsson caloric engines[7] and the equipment was installed at a cost of £619. On the deck immediately above the caloric engine, an air pump fed into a small sounding receiver only two feet high and about twelve inches in diameter. Set into the top of the receiver was the reed box and trumpet, extending to about three foot six inches. In case of mechanical problems, or simply whilst waiting for the engine to build up sufficient air pressure, two hand-operated air pumps could supplement or substitute for the machinery. It would have been hard work to sustain this for very long. If neither means of sounding the reed could be sustained the gong was resorted to, as the 1875 Admiralty List indicated: 'A gong or a fog horn is sounded at intervals of ten seconds'.

In October 1876, twenty-four of Mr Brock's rockets replaced the reed in trials but the reed was shortly restored to normal service.

It was usual to remove a lightvessel from its moorings for periodic repairs and maintenance and a substitute was towed into position whilst the former was out of service. For a variety of operational reasons, the original vessel was not always returned to its previous station and after repairs or refit the vessels were moved between stations as required. Some lightvessels did not remain at a particular station for very long, although there are also examples to the contrary.

Older lightvessels were gradually replaced but some of the early wooden boats remained in service for a long time: Number Two, built in 1825, was in service for eighty-nine years![8]

In 1879, another new vessel was placed on station at the Seven Stones and the 1880 List of Lights informed mariners that the fog signal on the new lightvessel was a 'syren' giving three quick blasts every two minutes. The automatic siren sounded at 40 psi and was powered by two 5 hp Brown caloric engines, similar to those installed at the Lizard in 1878, although smaller. Only one engine was usually worked, using about 18 lbs of coke per hour. The engines drove geared cams to regulate the siren blast. Including the air receivers, the configuration of the equipment occupied a central space below deck of 20 feet by 21 feet, which was the width of the boat. The trumpet protruded 12 feet above the deck house in front of the mast. This new ship was similar in size to its predecessors, although the deck house was larger.

The single lantern was 8 feet in diameter and a revolving mechanism produced a group flash of three every minute. The catoptric lantern had nine oil lamps fitted into silvered parabolic reflectors arranged in three groups of three. Above the lantern mast was the six-feet-diameter ball daymark.

---

[7]   The 1½ hp Ericsson engine had an 18-inch cylinder and ran at 45 rpm.

[8]   Lightvessel Number Seven, built in 1873, was sold after 99 years' service and an early iron vessel from 1857 lasted for 70 years. See Hague, D.B. and Christie, R., *Lighthouses, their Archaeology, History and Architecture*, Gomer Press, 1975. Lightvessel Number 95 was built in 1939 and is still in service after refits, automation and, most recently, the addition of solar panels to operate the light and the racon. Number 95 was on station at Seven Stones from 1978–1986.

Timber-framed, the vessel had teak decks and 4-inch thick teak planking. The 100-foot length of the hull was sheathed in copper and there were short keels to help stabilize the vessel. Bernard Waymouth of Lloyds designed the vessel, which was built at a cost of £9500. The sturdy construction guaranteed that vessels of this design survived in the service as late as the 1960s, although mainly as spares by that time.

Forward of the engine room was the coke bunker and other storage space beside the chain locker and mess room, and to the stern was the master and mate's cabin. Storage space for sails contained a lug sail, which could be hoisted on to the mainmast when the lantern was lowered. A mizzen sail was a permanent feature which steadied the vessel into the wind.

After sustaining serious storm damage in 1886, this lightvessel was replaced by another, Number 56, a vessel very similar to its predecessor except that it had iron beams rather than timber. The design and construction of lightvessels remained quite similar throughout the Sixties Class, which were built between 1888 and 1900, but after this time vessels were constructed of iron and, from the 1930s, steel.

Number 56 was fitted with a two-note siren with the characteristic low note, high note, low note blasts in quick succession every two minutes. A high- and low-note siren remained in use at Seven Stones until after World War Two, but in 1905 the character was changed to LHL every minute, coinciding with Number 56 being replaced.

Whilst most lightvessels had siren fog signals in the early 1900s, some reed fog signals were still in use, particularly on the older boats. The reeds produced a sound of about 500 Hz, although the 1901 trials at St Catherine's determined that 400 Hz was more satisfactory. The single directional trumpet had also been generally replaced by the vertical horn with mushroom resonator, although there were examples of multiple trumpets used to try and provide omnidirectional signals.[9]

The earliest reed fog signal used on a lightvessel was the Daboll trumpet installed on the Newarp lightvessel in 1868, but the Holmes-designed reed fog signal was used elsewhere, usually in combination with the Ericsson caloric engines as at Seven Stones.

The Barker reed, which was included in the 1901 St Catherine's trials, was used aboard some lightvessels. The throat of the trumpet contained three steel reeds and the fog signal used high pressure compressed air to sound a note of 350 Hz.

---

[9] The omnidirectional mushroom horn was introduced from the 1890s and although the design was always thought to be inefficient, it remained in service with the diaphone on lightvessels until the 1980s! An alternative to satisfy a 360 degree spread of sound, was to have three trumpets spread by 120 degrees. In Britain, this design usually had a reed in each trumpet, but there were examples where the trumpet was open to a single reed or siren. American lightvessels sometimes had three or four trumpets, resembling a clover leaf in plan, open to a single siren. If separate reeds or sirens were used in each horn some cancellation of the sound or 'beats' could result if the instruments were not synchronised.

Chance Brothers hand horn. Chance Archive, Courtesy Pilkingtons.

REED FOG SIGNAL. Actuated by hand wheel or treadle.
MANUFACTURED BY CHANCE BROTHERS AND CO., LIMITED, Lighthouse Works, near Birmingham.

Barker reeds were tested on the Girdler lightvessel in the 1890s.

Some lightvessels still relied on the gong, bell and rocket at this time, but in 1892 Trinity House established a sub committee to evaluate the adoption of a hand-pumped reed signal as an alternative to the percussion signals. The 'hand horn' was developed at Blackwall workshops and tested at the Nore lightvessel. It produced a note of about 440 Hz. Following the successful results of the trials, hand horns were made and supplied to twelve lightvessels, with a further three kept in store.[10] Similar hand horns were produced by other manufacturers, including Holmes, and these instruments were also used at small harbours and aboard sailing vessels. On lightvessels, the standard fog signal from a hand horn was two blasts in quick succession every two minutes, regardless of station.

The 'hand horn' reed was one and a half inches long and fitted into a brass holder which was screwed into the reed box on top of the receiver. It was sounded at about 5 psi. A pressure gauge and a clock assisted the operator and later versions could be pumped by hand or foot treadle. The blast lasted no longer than four seconds and the two blasts could drop the pressure to 3½ psi or less. The horn could be rotated into the wind and the interval between blasts could be increased should the signal be ignored by an approaching vessel. With gongs and bells and even rockets, if the approaching vessel failed to

---

[10]   Hand horns, made at Blackwall, were installed from 1893 at Nore, East Goodwin, Cross Sand*, Galloper, Gabbard*, Long Sand*, Swin Middle, Mouse, Prince's Channel, Gull, Nab and Cockle. Excepting the Nab, these vessels were all in the Thames approaches.

 * *There was a height restriction which required receivers to be four feet eight inches, all others were six feet three inches.*

alter course the frequency of the signal between intervals was increased to a continuous warning if necessary. However, the hand horn could only sound above a certain pressure, below which no sound of any worth could be forced out of it. At this stage, the carronade could be resorted to!

The manual reeds were intended to replace gongs but, in addition, found their way on to other lightvessels where they were used if the standard reed or siren was out of order, or if the engines were not ready. One of those used on the Seven Stones lightvessel is preserved in the Lighthouse Museum in Penzance.

Holmes' early design, from 1882, used gravity to drop a piston inside a combined cylinder and receiver. The heavy piston was slowly wound up by a geared chain and, when released, it fell and compressed the air, forcing it out through the reed box attached to the bottom. This reed was also used on some sailing vessels, as was a smaller and more compact device known as a Norwegian horn. This looked like a suitcase with a small horn mouth protruding slightly from one end and a cranking handle sticking out of the side. Winding the handle round produced a warbling wail of sound in the small reed. This device could also be found aboard many lightvessels.

Chance Brothers made a version of the manual reed horn which sounded at about 5 psi and which could be attached to a small 1 ½ hp oil engine in the deck house. In the event of a breakdown, the reed could still be manually operated at a lower pressure.

By the early 1900s, horizontal single cylinder oil engines had replaced most of the hot air engines previously used to operate fog signals at lighthouses and aboard

Holmes hand horn (left) and the Norwegian portable reed fog horn used by sailing vessels, lightships and small harbours. This one is at Penarth.

ightvessels. Trinity House typically used 9½ or 14 hp oil engines,[11] although 5 hp Hornsbys were used on lightvessels 32, 33, 38, 40 and 50, to power reed fog signals.[12] The 40-class lightvessels were generally refitted with 9½ hp engines from 1905 and lightvessels 51 to 79 were refitted or delivered new with 14 bhp engines. Lightvessel number 80, the last to be delivered to Trinity House before World War One, was built in 1914 and fitted with 22 bhp Hornsby engines. Number 80 later found itself placed at the Seven Stones.

The Hornsby engines ran on light mineral oil, otherwise known as 'lighthouse paraffin', which cost tuppence ha'penny a gallon at the turn of the century! The single cylinder Hornsby had a vaporizer at the end of the combustion chamber, connected to the cylinder by a narrow neck. A primus oil lamp was used to start the engine by heating the combustion chamber for ten to fifteen minutes until it glowed a dull red. The oil caps were filled and other lubricating points were attended to whilst the bulb warmed. When the vaporizer was hot enough and the air inside under sufficient pressure, oil squirted inside would cause an explosion. On smaller engines, the flywheel was manually pushed around to start the machine but on larger ones, the flywheel was over five feet in diameter and considerably heavier. Hence compressed air from the storage receivers was introduced to push the piston down the cylinder and pressurise air in the bulb on the return stroke. Air compressed in the cylinder was forced into the vaporizing chamber and near the end of the stroke a controlled amount of oil was injected into the hot bulb: the oil vaporized in the hot compressed air and exploded to force the piston out again. For the first few revolutions compressed air could be manually blasted in on-stroke to assist the engine. Once running properly, at about 220 rpm, the hot bulb remained hot and the primus lamp would be extinguished. The quantity of oil injected into the vaporizer was controlled by adjusting the governor, which also regulated the fuel pump allowing surplus oil to flow back into the tank in the base of the engine. A camshaft controlled the air intake and exhaust valves. The bulb was designed to stay hot but the cylinder had a water jacket and was cooled by circulating sea water on lightvessels.

Starting the engine was assisted by having already positioned the piston on-stroke by manually turning the flywheel. Failing this, starting the engine was possible by moving the flywheel in reverse, made easier with no compression and by skilfully admitting compressed air at precisely the right moment, so forcing the engine to run forward. This method was not recommended as the engine might continue to run in reverse, which would consequently fill the engine room with exhaust gases from the air inlet! Since I have heard this tale several times it probably happened more than once!

On the lightvessels, as at lighthouses, after the fog had lifted the engine and compressor continued working to fill the large air receivers to 120 psi ready for the

---

[11] In 1901 the most powerful engine Hornsby manufactured was 125 bhp.

[12] Lightvessels 32, 38 and 50 had triple reeds.

next time. This could take half an hour or more depending upon the receiver capacity and the engine size. Then, with the engines still warm, it was time to clean up. This included the floor, paintwork, brass, wiping down all the steel, mopping up all drips and splashes, cranking the flywheel to reposition the piston – everything had to be ready for the next time.

The two 14 bhp Hornsby engines aboard lightvessel Number 56 were installed in May 1905. Between 1880 and 1900, nearly twenty new lightvessels were built and existing ones were refurbished. From 1905, the first vessels were additionally equipped with a new navigation aid, the submarine bell.

Professor Tyndall calculated the velocity of sound in sea water at 20° C to be 4768 feet per second, more than four times faster than in air. Water is a good medium for transmitting sound, as Monsieur Colladon had proved in experiments on Lake Geneva in 1826. Some scientists believed that water density was more consistent than that of air so that sound in water would be less vulnerable to attenuation.

Proposals for developing submarine signals were raised in the 1860s, but years passed without much real experiment into their application. The development of electrical sound transmission led to experiments in America with underwater microphones at the end of the nineteenth century, but the technology lagged impractically behind the ideas.

Experiments nevertheless continued and correspondence between Trinity House and its Scientific Advisor, Lord Rayleigh, indicates that a submarine signalling system was proposed by Mr F.J. Ivens in August 1896. Rayleigh was not impressed by the plan, however: 'I cannot recommend the Elder Brethren to take up the matter … my own impression is that it would not succeed …'.[13] Rayleigh was much more interested in the potential of 'wireless electrical fog signals'.

In America, research into submarine signalling by Professor Gray and Mr Mundy was encouraged by the outbreak of war with Spain: it was thought that underwater microphones could detect enemy torpedo boats. Experiments used microphones which were suspended over the side of a boat, but it was difficult to overcome the problem of interference caused by 'ships noise'. Any signals which may have been picked up by the microphones were lost in the noises produced by the ship itself.

A major breakthrough occurred in 1903, when underwater microphones were abandoned in favour of receiving equipment inside the ship. Microphones were attached to small tanks on the inside of the hull below the waterline. The tanks contained a dense liquid solution, which communicated higher frequencies conducted through the water and the hull of the vessel, but screened out much of the lower

---

[13] Lord Rayleigh was not the only sceptic; the authors of an article in the *Syren and Shipping* journal in 1897 scoffed at the 'ludicrous pretensions' of the submarine signals. In Britain the Lightning Express Submarine Navigation Company Limited tried to raise £100,000 in a £1 share offer. Rayleigh, nevertheless, later came to accept the idea of submarine bells '… if it be true that they can be depended on …'. *Correspondence and Reports of the Trinity House Scientific Advisor.*

frequency sound made by and conducted through the vessel itself. Signals detected by the port and starboard receivers were linked to the wheelhouse, where the operator listened on headphones. Before long, unskilled listeners could not only distinguish a submarine signal above other noises, but specify its apparent position. Skilled operators could hear signals up to twelve miles from the vessel and correctly establish the bearing off with surprising accuracy.

Electrically-operated bells were struck under the water to produce the sound and experiments established that the design of the sound bow of a submarine bell had to be different to that of a normal bell in order to produce the best signal. Bells of about 140 lbs weight, 12 inches high and 14 inches in diameter, with a thickness of several inches, proved to be the most suitable, although some larger ones were also used.[14]

The Submarine Signal Company spent a lot of money developing the system and quickly established itself with shipping companies and the Canadian and US Lighthouse authorities. The reputation of the system was greatly enhanced by stories of ships navigating successfully in gale force blizzards when it was impossible to hear a fog signal or see a light. By 1905, the first submarine bells were being installed on lightvessels for tests in this country.

The submarine bell was lowered over the side of the lightvessel on davits and the striker was operated by compressed air. The fog signal actuating clock could sound the bell with whatever characteristic was required. On some vessels the bell was permanently fixed below the water line, as it was on automatic light floats and on those buoys which were fitted with a submarine signal. Submarine bells used on buoys were usually wave actuated, such as those placed near Eddystone and Bishop Rock lighthouses.

Many lightvessels had the new bells by 1907 but the installation programme continued into the 1920s and a few bells were additionally placed on the sea bed near lighthouses, operated remotely by compressed air or electric cable. A submarine bell was installed near South Stack lighthouse in 1907 and at the Lizard in 1910. The Seven Stones lightvessel was fitted with a submarine bell in November 1913.

From the mid 1920s, the submarine bell was superseded by the submarine oscillator. This signal used the same principle as the bell but developed the advantages of electrical synchronization with radio transmissions and gave a sustained note for a longer period than the single stroke of the bell. Like the bells, the oscillator was lowered over the side of the lightvessel and once submerged, produced an audible note by means of the electromagnetic oscillation of a metal diaphragm at about 1050 Hz. These vibrations must have been popular with the unfortunate lightshipmen!

---

[14]   Striking the bell in the denser medium of water required a greater expenditure of energy than in air and to account for the difference in sound transmission, Rayleigh suggested tuning the bells to 'C', or preferably a semitone higher (1088 Hz). *Correspondence*, ibid.

Sounds from the bells and oscillators could be detected through a ship's hull at short range, but new receiving equipment was necessary for the oscillators to be properly utilised. The sound range was greatly improved, as oscillators could be detected at fifty miles, compared to the bell which had been limited to about fifteen. As with the bell, the oscillator enabled safe navigation in fog, at least as far as fixed obstacles were concerned. Other moving vessels still posed the same dangers as before, although a proposal made by the Submarine Signal Company in 1900 was to fit the new bells on to ships. The idea was that the ships would then be able to indicate their positions to other vessels as well as calculate their own position from fixed signals.

The skills required of a lightshipman steadily changed from those of lamp trimmer and carpenter to those of mechanic, electrical engineer and radio technician; but much of the work remained very physical and little changed as far as cleaning, lifting, storing, checking, polishing and front-line maintenance were concerned. Some of these jobs had to be done in the exposed cold and wet conditions on deck, including keeping watch at night. Even off-watch, the noisy oscillator and fog signal carried on as before!

The oil lantern still had a good long life ahead of it, although electricity became increasingly used aboard the lightvessels. After the First World War, seven years passed before the next new lightvessel was built. The Armstrong Whitworth shipyard in Newcastle built four new ships and the first, Number 81, was completed in 1925. These new lightvessels were 104 feet long with a beam of 25 feet and had electric lighting. All four ships were fitted with diaphone fog signals and the compressors and generators were driven by 72 bhp Gardner T engines, similar to those described at Lundy North lighthouse. The lightvessels used four cylinder engines. The lantern was fixed on top of a tubular tower, which enclosed the steps up to the lantern and the optic was gimballed to move independently of the lightvessel's own rolling on the waves. The tower was not amidships but a little to the stern. These lightvessels were all equipped with submarine oscillators.

When all four of the new ships were completed in 1930, Trinity House had 46 lightvessels on station with an additional eight spare, although half of the total number then in service had been built before 1880.

Number 81 was to be placed at the Royal Sovereign and the 1924 List of Lights anticipated this occurring in 1926, promising that a diaphone would replace the siren and an oscillator would replace the 1907 submarine bell. Number 81 had a K-type diaphone and even though the Royal Sovereign station was seven miles off the coast and the previous vessel had sounded its two-note siren for many years, the residents of Eastbourne found the new fog signal more than they could bear, or so the story goes! Trinity House obligingly replaced the vessel and Number 81 went off to the Kentish Knock where the noise caused less distress, except to the lightshipmen!

The diaphone was tested in England and Scotland before World War One and was produced in Britain in the 1920s. (Chapter 11 covers the diaphone in more detail.) A diaphone was in service at Casquets lighthouse in 1921, at Barrow Deep lightvessel in 1925 and at Outer Gabbard lightvessel in 1924, along with an oscillator. The combination of diaphone and oscillator was also proposed for Smiths Knoll but not installed, and in 1927 Newarp lightvessel had a diaphone and oscillator.

The basic operation of the submarine oscillator was not very different from that of the early electric horns or emitters, which were often called 'air oscillators', and which were developed and introduced in the late 1920s and early 1930s. Electric signals were developed largely by the Americans and Germans and some emitters found their way into service, primarily at harbours, as their range was not comparable with the diaphone or siren. For this reason they were not used on lightvessels,[15] even though the idea of synchronized beacon, oscillator and electric fog signal had been considered.

The earliest electric emitters in service in Britain were at Swansea in 1925, followed by Southend, Fawley and Bridlington. Other 'nautophones' followed on the east coast and at Scottish harbours and fishing ports in the 1930s, but at several of these sites air signals were reintroduced in subsequent years.

The US Lighthouse Board was favourably impressed by the performance of the 'air oscillators' and in trials in the 1930s the range of signals tested compared reasonably well with compressed air fog signals. Significantly, the oscillators used considerably less energy in producing the sound in comparison with the sirens and diaphones, which required large engines to supply huge quantities of compressed air.

Chance Brothers manufactured an electric oscillator from 1939 which produced a low note of about 100 Hz. The diaphragm was large, half an inch thick by twenty inches in diameter. (Some diaphragms were much thicker.) The resonator designed for this diaphragm used the oval mouth principle developed by Rayleigh. A 300 Hz diaphragm signal was also experimentally produced by Chance Brothers but design problems caused it to use excessive energy and it was not developed any further. Later developments of electric fog signals are examined in Chapter 13.

As previously noted, radio beacons were significantly spread in the 1930s and certain lightvessels formed part of the network. Experimental, synchronised radio beacon and fog warning signals were also trialled at some lightvessels, such as Breaksea in the Bristol Channel. A Notice to Mariners issued in 1938 indicated that a synchronous fog signal would supplement the siren and radio beacon. The beacon character at the time was one transmission of the call sign GGA repeated three times and followed by an interval. This was followed by a nine second dash and another interval before the 'warning signal' – a single dash of two seconds – then twelve synchronous dashes lasting 16 seconds. This transmission ended with the call sign and the whole cycle was repeated every six minutes. The submarine

---

[15]   The Spurn lightvessel had an electric emitter in 1946. The light vessels stationed at Spurn were built for and operated by the Humber Conservancy Board.

oscillator, meanwhile, sounded the Morse letter A (dot, dash) three times in quick succession every 30 seconds:

> *'The first and third transmission in every twelve of the Submarine Fog Signal begins at the instant the 'Warning Signal' of the Radio Beacon ceases, so that two synchronized signals with an interval of one minute between them will be sent out every six minutes during fog … The number of dashes heard[16] between the 'Warning Signal' and the first note of the Submarine Fog Signal indicates the number of miles the observer is distant from the lightvessel'.*

At Nab Tower off the Isle of Wight, a synchronized diaphone and beacon signal was established in 1938. Here a stopwatch was required by the officer on the bridge, which was to be started on hearing the commencement of the second long dash on the radio. The watch was stopped when the diaphone was heard and the number of seconds elapsed was multiplied by 1.87 to give the mariner the vessel's distance, in cables, from the Nab. The Notice to Mariners stressed that 'The speed of sound in air varies with temperature and barometric pressure … and the diaphone blast and not its echo must be the reference'. The radio beacon signal contained two long dashes and it was the second one which must be used. Trinity House requested that observations on the possible value of 'Experimental Synchronised Signals should be sent to the Secretary at Tower Hill'.

Whilst such aids to navigation were still being evaluated, some lighthouse authorities had already recognised that submarine signals had outlived their usefulness by comparison with the development of the radio beacon network, which provided all of the information that could be obtained from the submarine signal.[17] However, supporters of the submarine signal believed it was the best possible navigation system and were very disappointed that it was being squandered by ship owners. The ship owners were equipping their vessels with radio receiving and direction finding equipment and were reluctant to duplicate this with submarine signal receiving equipment.

In Britain, submarine signals were suspended during the war years and the synchronized 'wireless and submarine sound signals' transmitted from lightvessels were abandoned altogether soon after the war, although charts continued to show 'obsolete' submarine bells and oscillators until well into the 1950s.

During World War Two, the use of lightvessels and lighthouses was restricted to timetabled assistance to convoys and many lightvessels were taken off station. The exposed positions of the remaining lightvessels left them vulnerable and many came under attack, not just those in the Thames approaches. Two vessels were sunk by enemy action, including Number 69 at South Goodwin and after these incidents,

---

[16]   The dashes were one and one fifth of a second apart.

[17]   The Canadian authorities had decided to develop the radio beacon network and were a little surprised at the strength of interest in submarine signals retained by European lighthouse authorities in the early 1930s.

some crews were taken off between convoys. Most lightvessels had remained on station during the First World War but had been under the command of the Royal Navy for the duration.[18] Other war service saw Trinity House lightvessels marking safe channels for the D-Day landings in 1944.

The Seven Stones lightvessel was moored in the River Fal for much of the Second World War and towed back to its station afterwards. This was Number 80, already over thirty years old. Besides the class Eighty and Ninety lightvessels, many old wooden vessels were repositioned after the war, although some of them urgently needed major refits or replacement.

Phillips of Dartmouth, who had built the Nineties class lightvessels from 1936, resumed work for Trinity House after the war. Number 95 was the last vessel built by Phillips prior to World War Two and is currently in service at Inner Dowsing. It is one of the last ten lightvessels still in service, although all of them are now fully automatic. Number 95 was stationed at the Seven Stones in 1978.

A total of 22 steel lightvessels were built for Trinity House after the war, most of them quickly introduced in the 1950s. The numbering of the lightvessels began again after the war and numbers 1, 2 and 3 were completed in 1946. The new vessels were equipped with three or four cylinder Gardner LW diesel engines and all had electric lights and diaphone fog signals.

The war years hastened the development of radio navigation and radar and radio beacons were improved after the war. Call signs, frequencies and transmission times were internationally agreed in 1951. Radar developments led to experimental trials with the 'RAdar beaCON' such as at the Tongue lightvessel from 1953: 'The Racon flash is a greatly lengthened and reinforced echo'.

The experimental racon service was discontinued for appraisal in the following year at the Tongue lightvessel, but the racon has been a standard feature not only on lightvessels but at lighthouses and on larger buoys. A ship's radar triggers a response from the racon, which gives a visual indication on the ship's radar display indicating the range, bearing and identification of the beacon – a Morse characteristic. On the lightvessels the racon transmitter was attached to the lantern gallery or on top of the lantern.

The Phillips post-war lightvessels usually had open lattice towers amidships although the later vessels had enclosed towers. Dioptric lights had been used on some of the pre-war vessels and lenses made by Chance Brothers had panels which could be adjusted or inverted to enable the required character to be shown. The post-war lightvessels had catoptric lights and although the groups of silvered

---

[18]   The Admiralty had previously intervened during the Napoleonic Wars to block the placing of the Sunk lightvessel which they believed would assist the enemy. To safeguard the fleet, however, the Admiralty insisted on placing lightvessels at Galloper, 1803, Gull, 1809 and Nab, 1812 and Trinity House subsequently retained these signals.

The 'mushroom' omnidirectional trumpet.

reflectors were now illuminated by electricity, the principle was unchanged from those reflectors introduced over a century before!

Forward of the tower was the diaphone, which was installed atop the wheelhouse. The 90s class had impressive wheelhouses, but the term is rather a misnomer when applied to the post-war vessels – there was no wheel! Inside, there was instead the flag store and the operating and sounding admission valves for the diaphone. Air pipes climbed the walls and ran across the roof to the bottom of the diaphone case. The rest of the diaphone and its cast iron resonator extended vertically out through the roof and culminated in the mushroom reflector.

Climbing down into the deckhouse one would find two air receivers, the reducing valve, the fog signal coders and the stop valves which opened the receivers to the diaphone. Another stop valve opened the air supply to the capstan on the deck outside, which powered a winch as well as tightening or slackening the mooring chain as necessary.[19] Large diameter pipework with gentle bends covered the wall space and a door on the far right led into the engineers' store. Steep steps led down into the engine room.

---

[19] If the service tender was delivering supplies whilst the diaphone was sounding, it apparently encouraged the crew to work extra hard: loading coal with the air-driven hoist could reduce the air supply to the diaphone!

Flanking the steps in the engine room stood two large upright air receivers and in the middle of the room was a cast iron stove, important to prevent cooling water freezing in the engines. On either side of the stove and the narrow gangway to the steps stood the Gardner 4LW diesel engines which produced electricity and compressed air. Trinity House lightvessels had no motive power and had to be towed to and from their stations. Each engine was attached to an electric motor and by Vee belts to the Reavell Quadruplex compressor. Extra safety guards distinguished the engine and compressor set from its equivalent ashore; when the lightvessel was lurching about, the exposed parts were obviously a greater danger to life and limb.

When the fog signal engine was not operating, one or more of the four single cylinder Gardner engines were used for generating electricity. These engines, first made in 1929, were aboard most of the post-war lightvessels and they were also used at certain lighthouses. The electric alternators were later made by Crompton but earlier engines had a combined Gardner dynamo. The IL2 engines ran on paraffin and developed 11 bhp at 1100 rpm.

Black Bakelite and brass switching equipment with rheostats and knife switches was on the far bulkhead of the lightvessel engine room.

The hull forward of the engine room contained coal, oil and water stores and the chain locker; towards the stern of the vessel was the mess room, crew's quarters and master's cabin. The deckhouse above contained more stores, the showers, heads and galley and the watchroom, which contained the communication equipment and the auxiliary fog signal coder.

From the early 1960s, an auxiliary, electric fog signal was installed aboard the lightvessels at the base of the tower. These signals were battery-powered and intended for use if the diaphone was not operational. Four Tannoy 100 watt pressure units, fitted inside protective conical casings, were assembled into a vertical column with the speakers facing downwards. Large, powerful, electric fog horns were experimentally used as the main signal on one or two vessels, including Number 88.

Off watch, on a beautiful, calm, clear day with fresh sea air and cooking smells coming from the galley, maybe fishing or just relaxing and anticipating the boat relief, life probably felt great! In the winter cold, dark and wet with nauseating diesel vapours in the bilges and with frozen hands aching to free jammed machinery, with the vessel rolling horribly in a wild sea and with the diaphone curdling the blood three times every minute, life probably felt …! Some lightshipmen wondered why they did the job; others could not wait to complete their one and only tour of duty.

The working environment could be unpleasant and dangerous aboard the lightvessel. Minor engine room accidents could be supplemented by slipping or falling and occasionally, a man fell overboard. Disturbingly, some lightvessels and their crews suffered worse accidents, namely collisions with other vessels.

It was not impossible for a lightvessel to drift off station and in 1954 seven lightshipmen were killed when Number 90 broke loose from its moorings in a storm

The standby electric fog signal in use from the 1960s.

and was driven onto the South Goodwin Sands. Six years earlier, Number 80, at the Seven Stones, drifted off station to end up 13 miles north-west of Pendeen lighthouse, although fortunately without tragic consequences.

Number 80 was refitted in 1954, when a diaphone was installed, and remained in service at the Seven Stones until finally replaced by a new lightvessel, in 1958. This was Number 19, which had an enclosed tower and a diaphone turret installed towards the stern of the vessel rather than in the more usual forward position.[20] The G-type diaphone gave three blasts every 60 seconds during fog.

Originally fitted with masts fore and aft, the stern masts were removed in the 1970s to accommodate the installation of helidecks. Despite the diaphone turrets, the later vessels also had helidecks fitted and from 1978 Sevenstones was serviced by helicopter.

[20] Lightvessels 18– 22 differed from the earlier post-war vessels in appearance – most obviously in the tubular rather than open lattice tower and stern-mounted diaphone. The later vessels operated on AC generated by six cylinder Ruston diesels where the earlier ones used DC. The last lightvessel ordered by Trinity House, Number 22, was built by Richards, and launched at Lowestoft in April 1967. Trinity House later purchased three secondhand vessels from the Humber Conservancy Board, the Mersey Dock & Harbour Company and the Commissioners of Irish Lights. Thus Number 23 was actually launched in Dartmouth seven years before Number 22.

Seven Stones lightvessel in 1952 when No. 80 was on station. The diaphone was the standard fog signal on new lightvessels from No. 81 onwards. No. 80 was built in 1914 and retained its three siren trumpets until replaced by a diaphone in 1954. Courtesy Trinity House National Lighthouse Centre.

Seven man crews worked the lightvessels until the encroachment of automatic systems resulted in a reduction to five men, followed by the complete de-manning of the remaining vessels in the late 1980s. The very last crew was taken off the Channel Lightvessel, Number 23, in 1988.

In the 1970s, it had seemed that 'Large Automatic Navigation BuoYs' would replace lightvessels but this did not finally materialise. Lanbys were used in this country and abroad and their automatic operation was remotely monitored. Inside the Lanby were diesel engines which ran continuously to power the light, electric fog signal, communications and racon.

An essential problem with the Lanby, however, was that it did not ride the sea like a lightvessel and its forty-foot diameter shallow bottom caused it to lurch violently on the waves. By and large, the equipment tolerated most of this treatment but when an engineer's attention was required, it was very difficult to get him on to the Lanby, except in calm conditions. Even if he could get aboard, the discomfort suffered guaranteed that he couldn't get off it soon enough!

The automatic light float by comparison rides the sea in a comparable way to the lightvessel. As noted in Chapter 1, automatic light floats have a long history. In the early days, the lantern was Pintsch gas or acetylene operated and gas was often also used to operate the fog bell. The automatic light float was not always a satisfactory solution, however, and if a first-class fog signal was required, there was no alternative but to man the lightvessel so that fog signal engines could be operated and maintained.

Modern light floats are solar-powered and the remaining automatic lightvessels are all likely to be converted to solar power in the near future. For the time being, these vessels will retain standard two or three mile electric fog signals sounding at 800 Hz.[21] Automatic lightvessels are currently in service at ten stations[22] with light floats at four[23] and all are remotely monitored at Harwich.

Many former lightvessel stations have been discontinued altogether; at some there are now only buoys. It is possible that other remaining automatic lightvessels and light floats will subsequently be replaced by buoys.

The High Focal Plane buoy has been in service for many decades but the limited range of the light and low height of the focal plane above the sea meant that the HFP could not compare with the much greater range of the lightvessel. It was also impossible to provide a first-class fog signal except on a lightvessel.

In recent years, the range of the racon and its effective operation by solar power has established the HFP buoy as a satisfactory and much more economical aid to navigation than the automatic lightvessel. Category 1 buoys sometimes have wave-actuated whistles but otherwise have no fog signal.

The HFP buoy has not proved suitable for all stations, however, especially at exposed sites such as the Seven Stones, where a trial buoy was installed but subjected to violent forces in the huge seas. An automatic lightvessel, Number 21, is currently still in service at the Seven Stones, so that with only a few interruptions, a little red boat has been on station here for over 150 years.

The master of the first Seven Stones lightvessel thought the boat rode the waves well and was '… always ready for it …', even when the deck was heavily washed over. Every time the sea struck the bow, however, it sounded '… like a four pounder going off …'![24]

---

[21]   ALV Number 21 has a 500 Hz emitter with an SA 850 reserve signal.

[22]   Inner Dowsing, Newarp, Sunk, F3, East Goodwin, South Goodwin, Sandettie, Greenwich, Channel and Seven Stones.

[23]   Varne, Owers, Breaksea and Bar.

[30]   *Penzance Gazette*, August, 1841.

For an enthralling and rare insight into life aboard lightships around the world see *No Port in a Storm* by Bob MacAlindin, Whittles Publishing, ISBN 1-870325-37-0.

Chapter 11

Portland Bill

and the diaphone

The two lights built on the Isle of Portland in the early eighteenth century were originally lit by coal. As leading lights, they warned of two main hazards off Portland Bill, the Shambles sand bank and the race caused by the tides running between the shore and the shoal. The original lights had no fog signals but Portland Harbour breakwater lighthouse had a bell from 1840 and the Shambles lightvessel, placed in 1876, originally sounded a reed which was later replaced by a two-note siren. From 1947, the lightvessel had a diaphone. An explosive fog signal on the breakwater was replaced by an electric fog signal in the late 1960s.

The movement of the Shambles lightvessel was a useful indicator of the state of the tide and the race, swinging to face E, NE on a falling tide, and W, SW on a rising tide. From two miles east of the Bill, the Shambles extends for over two miles. Heavy seas still break on the Shambles but the hazard is now marked by two buoys. The race remains a danger to sailing vessels.

The two old light towers still stand but a new lighthouse was completed on the end of the Bill in 1905, one of the tallest towers built in Britain at 136 feet high. The optic is a first order lens with four asymmetrical catadioptric panels and a dioptric mirror, and the whole apparatus makes one clockwise revolution every 20 seconds to produce the characteristic group flash of four. The clockwork drive was replaced on electrification and a 3 kw filament lamp replaced the pressurised vapour burner. A mercury vapour lamp later replaced the filament lamp. A subsidiary red sector light protects an area over the Shambles.

The tall tower contains only two floors between the service room and the lantern and the large space echoes with every voice, footstep and any other sound. From the base of the tower, the stairs spiral clockwise up to the first floor.

When the lighthouse was still manned, there were two sets of Reavell compressors attached to BTH induction motors in the base of the tower for the operation of the fog signal. One compressor set was used with the other on standby, and in the event

Portland Bill lighthouse and diaphone trumpet.

of a mains power cut there was a standby generator situated behind the second compressor set. On the right-hand side were three pairs of superimposed vertical air receivers, 6 feet by 3 feet in diameter, which could all be connected by opening stop valves.

Running at 955 rpm, the Reavell Quadruplex compressor was capable of delivering air at 150 psi. At 150 psi, each receiver had a capacity of 38 cubic feet of stored air but the pressure was normally maintained at 45 psi and only four of the six receivers were used. About 35 feet of six-inch diameter pipe supplied air from the storage receivers up through the tower to the sounding receivers on the second floor. The stairs spiralled around the receivers to give access to the iron platform of the fog signal gallery situated a few feet above the second floor.

The two sounding receivers were vertically stacked like the storage receivers in the base of the tower, only smaller, each one being four feet high and two feet in diameter. The air main fed into both receivers through reducing valves set at 35 psi, the operating pressure for the fog signal. The upper receiver connected to the 4-inch admission valve and the lower receiver supplied air to the 2½-inch operating valve and the coder, which was situated between the receiver and the diaphone. The small admission valve opened to admit air to the rear of the diaphone and the larger admission valve supplied air to the sounding chamber of the diaphone.

Portland Bill had an F-type diaphone positioned in a former window opening in the tower about 65 feet above the ground. The slight bell mouth of the cast iron resonator protruded a little way out of the tower. Before the installation of the diaphone in 1940, Portland Bill had no fog signal.

The diaphone was discontinued as part of the automation of Portland Bill lighthouse in 1995 and Trinity House has installed a high frequency electric fog signal and a fog detector, which automatically starts the signal when necessary. The keeper, on the other hand, used to look up and down the coast from the lantern, or possibly stand by the water's edge watching the movement of the fog banks slowly rolling in off the sea, accounting for the speed and direction of any wind and judging the visibility by eye. When it was obvious that the fog was bound to arrive shortly and fog marks drifted into the murk, the keeper started the signal. Paragraph 37 in the Lighthouse Service Regulations gave a clear direction to keepers: the '... fog signal is to be sounded too much rather than too little', and a 1921 Trinity House circular added that

> '... keepers are instructed that as the fog signal is intended not only for the benefit of vessels passing close to their station, but also as an aid to navigation in the offing, it should be sounded when there is the slightest indication of fog'.

At night, the keeper looked for familiar lights at known distances to gauge visibility. The arrival of fog was also apparent through noticeable changes in sound clarity and apparent changes in the lighthouse beams, which became more conspicuous and yellow and penetrated less far, appearing to 'drop' towards the ground.

Having decided to start the fog signal, the keeper checked the compressor oil levels and the electric motor was switched on. The clutch between the motor and the compressor was then engaged by winding in the wheel, at which point the noise level in the towers would dramatically increase as both of the unloading valves on the Reavell noisily chattered. After winding the valves in,[1] the noise from the compressor was reduced, but the inlet valve on the storage receiver needed to be opened fairly quickly. The stop valves on top of the receivers were then opened to supply air to the sounding receivers upstairs. If the gauge indicated the wrong pressure in the storage receivers, the relief valve on the compressor was adjusted. This valve regulated the output from the compressor and dumped excess air to the outside.

---

[1] The engine or motor was started off-load and the clutch engaged automatically by hydraulic friction or, as at Portland Bill, manually wound in. With the compressor on load the receiver stop valves were opened and the two unloading valves on the compressor closed.

With everything running smoothly down in the base of the tower, the keeper climbed the stairs to the fog signal gallery, started the coder and gave the pressure gauge on the sounding tanks a cursory glance so that it was obvious if he had not opened the outlet valves on the storage receivers! Within thirty seconds, the squeaking and groaning of the compressed air in the receivers and the gentle buzzing of the coder motor would all be swamped by a thump and the sound of escaping air, followed, almost immediately, by a heavier thump and a terrible noise which reverberated and resonated throughout the effective sound box of the tower!

The 3½-second blast produced a note of about 180 Hz during which the pressure gauge indicated a fall of about 4 psi. The diaphone sounded once every 30 seconds and the published audible range was seven miles, but this would vary according to conditions. Under favourable conditions it could be double that.

The diaphone produced a distinctive 'grunt' at the end of each blast which effectively increased the output of sound by up to 8 dB and at the maximum range of audibility, it was the 'grunt' which was invariably heard by the mariner. The similar effectiveness of the 'whoop' at the commencement of the five-inch siren blast has already been noted. The 'grunt' was a controllable feature of the design of the diaphone and effectively distinguished the signal from every other type.

Whilst the fog prevailed, the signal could be left to itself with periodic checks on pressure and temperature gauges. Shutting the signal down afterwards was the reverse of the start-up procedure, except for leaving the coder running until last to empty the sounding receiver.

The fog signal equipment required periodic inspection, initially by the keepers.[2] Bolted to the rear of the black cone of the resonator was the diaphone. Inside this unimpressive looking cast iron casing was a gunmetal cylinder, containing a hollow piston with sounding ports. The casing had two annular spaces, one which surrounded the sounding ports along the length of the cylinder and the other at the cylinder head, into which the operating, or driving, air was admitted.

When the piston was set oscillating by the driving air, the rapid vibrations caused by the escaping air through the sounding ports produced the sound in the resonator.

The standard length of the cast iron resonator for an F-type diaphone was 4 feet 2¼ inches.[3] The Canadian lighthouse authority referred to tuned resonators, but a Chance Brothers booklet, published before the company began diaphone manufacturing themselves, said: 'One advantage claimed for the 'Diaphone' is that it automatically adjusts its note to that of the trumpet'.

---

[2] Service regulations instructed keepers to pay particular attention to the diaphragms on the admission valves, which required replacement in any case every twelve months. Rust blown through the system could be a particular nuisance and could cause damage to valves and the diaphone or siren itself. Air cleaning filters were installed in systems from 1960.

[3] Resonator sizes: 'A-type' 2 feet 3 inches long, 5¾-inch diameter mouth. 'F-type' 4 feet 2¾ inches long, 16½-inch diameter mouth. 'K-type' 5 feet 3 inches long, 22 -inch diameter mouth.

Diaphones were produced in different sizes and all were designed to sound at about 180 Hz, double the frequency of the piston cycle, but because each instrument was slightly different, each produced its own specific note somewhere between 180 and 200 Hz.

The characteristic 'grunt' at the end of the blast was produced by stopping the operating air to the diaphone before the sounding air, with the consequential stalling of the piston inside the cylinder. This produced a tumultuous drop in the fundamental note, harmonics and beats as the oscillations stopped.

The diaphone at Portland Bill was controlled by the coder which dated from the 1960s. There were actually two electric coders, A and B (one was a standby) and the equipment was produced at Blackwall workshops. Many similar ones were installed at fog signals in the 1960s and 1970s and the electric coders fulfilled the same function as the clockwork actuating machinery they replaced. A geared electric motor accurately rotated a disc with notches in the circumference which corresponded to the duration of the fog signal blast and interval. As the disc revolved, electrical contacts first tripped a solenoid actuating valve to open the control air supply to the operating, or driving, valve and secondly to supply control air to the main air admission valve.

The admission valve was similar to those previously described. A flexible diaphragm, usually rubber, was forced by the control air pressure from the actuating valve against a large bore piston surface inside the admission valve. This moved the valve stem against a spring to open the valve. Whilst current to the solenoid was maintained, the actuating valve sustained the control air pressure on the diaphragm which kept the admission valve open. When the coder shut off the control air to the admission valve, the air pressure behind the diaphragm was released and the valve spring and opposing air pressure in the main pipe slammed the valve shut.

The coder first opened the small admission valve to allow the operating air into the diaphone casing and then the cylinder head, through a circumferential ring of small holes. This escaping air acted on the front of the driving head to push the piston backwards. This motion covered the inlet ports and exposed another ring of ports which admitted air to the back of the head. This pushed the piston forwards, allowed the air to escape and re-opened the original ports. This reciprocating cycle started instantaneously and was maintained for as long as the operating air supply was open. The stroke of the piston was 0.3 inch and the back and forth oscillating cycle was completed between 90 and 100 times per second. The coder opened the operating valve slightly before the sounding valve.

The sounding portion of the F-type piston was 3.125 inches long and had nine circumferential slits cut into the walls, each 0.048 inches wide. There were identical slits in the cylinder wall to correspond with the piston slits and these coincided twice on every stroke, rapidly opening and closing with the reciprocating piston movements. When the coder opened the sounding

Left: A G-type diaphone. The main air valve is at the bottom and the 'operating' valve is at the top. When the operating valve is opened the piston starts oscillating and when the main air valve opens the diaphone sounds.
Right: A spare diaphone cylinder and piston. The piston on the right fits into the cylinder, and air escaping through the slits in the wall produces the sound as the piston oscillates back and forth.

valve, the air let into the annular space surrounding the cylinder escaped in bursts through these slits and out through the open end of the piston, causing rapid compression and rarefaction of the air in the throat of the resonator, to produce the sound.

During the nineteenth century, there was an upsurge in organ building and the diaphone was originally designed for use as an organ stop. It provided a means of producing a controlled vibration to boost the strength of low notes. A 1913 book on organ building proclaimed that:

> *'The invention of the diaphone by Hope Jones in 1894 will someday be regarded as the most important step in the advance hitherto achieved in the art of organ building'.*[4]

No reed pipe could compare with the diaphone:

> *'By no other means known today can anything approaching such a grand and dignified Diapason tone be produced'.*

[4] Miller, G.L. *The Recent Revolution in Organ Building*, New York, 1913.

Robert Hope Jones was an amateur chorister and organist who was professionally qualified as an engineering draughtsman. He left the Laird shipyard in Birkenhead to join the new Lancashire & Cheshire Telephone Company in 1881, where he finally became chief electrician.

Hope Jones brought his skills and interests together to voluntarily repair and rebuild a church organ before he had left the shipyard and by 1889, he had established a company to rebuild and manufacture organs. His company developed electro-pneumatic mechanisms which drew on the technology of the telephone, and astute business practice established links with traditional organ manufacturers and repairers. The company secured contracts for repairing and restoring many church organs and those in public buildings and by 1894, Hope Jones had been commissioned to reconstruct two organs in Worcester Cathedral to make a single instrument. One of these organs dated from 1666 and Hope Jones has subsequently been seen as a vandal who destroyed original instruments and replaced them with primitive electrical machinery which lasted only a generation at best. His work has been dismissed as that of a technician, not a musician. His widespread work has itself now mostly been replaced, although parts of the Worcester Cathedral organ have survived, but are unused.

Hope Jones installed a remote electric console which combined the pipes of both original organs at Worcester, but he replaced many of the original stops with his own inventions, including the diaphone. The two diaphone pipes were operated by the pedal organ and were respectively 16 feet and 32 feet long. The diaphones at Worcester had a bellows attachment on a wooden box on the organ pipe. Spring-loaded valves closed the pipe until the pressure of air admitted to the box closed the bellows and caused the valve to open. Some of the air in the pipe was then directed to the rear of the bellows which filled and thus closed the valve. With the valve closed, the air supply to the bellows was also cut off and so they collapsed, which opened the valve again. This cycle was repeated for as long as the organist wished and the note was produced by the compression and rarefaction of the air in the pipe.

The same principle, but of a more elegant design, was incorporated into the diaphone for the organ at Aberdeen University. Other variations used electric motors to start the vibrations. Hope Jones patented his development and also registered a fog horn patent in 1896.

By 1903, his business difficulties had resulted in close calls with bankruptcy and he left for America to escape legal action. With a fresh start, he established the Hope Jones Organ Company of New York in 1905, but within five years this company too was bankrupt. By this time Hope Jones had developed his main interest, the electric organ, and when the Wurlitzer Company acquired his patents and designs, he was appointed as a division manager. Further financial difficulties drove Hope Jones to suicide in 1914, but Wurlitzer organs flourished in the new theatres and cinemas.

The diaphone installed in the organ at St Patrick's Cathedral in New York in 1905 had a reciprocating aluminium piston in a cylinder with corresponding slits cut in the walls of both. The piston did not have a flange but was sealed towards the bottom. Air entering the operating port at the bottom forced the piston upwards to uncover sounding holes; pressure above the piston then returned it to its starting point, uncovering the operating air ports again. The escaping compressed air followed by the corresponding rarefaction of air, produced a sound wave with a frequency determined by the length of the organ pipe.

Hope Jones' diaphone designs were acquired by J.P. Northey who patented the use of the diaphone as a fog signal in 1902. Northey owned a plumbing and pipe manufacturing company in Toronto and established the Diaphone Signal Company Limited in 1902. The company developed the diaphone and introduced seven different sizes of the instrument over a period of a few years. Northey's designs improved the efficiency of the diaphone and reduced the sounding pressure from 90 psi to about 30 psi. By 1903, the Canadian Department of Marine and Fisheries had been sufficiently impressed by the diaphone to adopt it as the standard fog signal at Canadian lighthouses.

In 1906 Northey approached Trinity House with the diaphone and Lord Rayleigh and Thomas Matthews considered its possible advantages over the siren. One of the disadvantages appears to have been that the diaphone used much greater quantities of air than the service five-inch siren, with which comparative tests were made. Rayleigh submitted a report to the Trinity House Board in July 1906, in which he described the diaphone and its 8½-inch diameter cylinder, but noted the greater air consumption which was five times more than the siren:

> '*I ... suggested to Mr Northey that he construct a five-inch diaphone, to consume as nearly as possible the same quantity of air as the Trinity House siren (working at a pressure of twenty-five lbs)...'.*[5]

At St Catherine's lighthouse in May 1907, comparative trials between the five-inch siren and a five-inch diaphone, both sounded at 25 psi, produced results in favour of the diaphone. The tests continued with a new design of horn made at Blackwall workshops, intended to further improve the diaphone. Horns tested included a Rayleigh trumpet and the French-made brass horn, previously used at the 1901 St Catherine's fog signal trial.

Matthews thought that Northey's price of £947 for the diaphone was very expensive; its merits did not exceed those of the five-inch siren to that extent! Northey also demonstrated the diaphone to the Northern Lighthouse Board at Inchkeith, where it was compared with the two-note, seven-inch Scottish siren.

The Canadian lighthouse authority believed that the diaphone was economical in use when overall results with other fog signals were compared – by 1909 they had nineteen in service on the St Lawrence River alone.

---

[5] Unpublished correspondence between Trinity House and the Scientific Advisor.

'All our diaphones are designed to give the note of F Sharp in the Middle Register, having one hundred and eighty vibrations per second'.[6]

The Canadian diaphones were generally F-types, similar to the one at Portland Bill, but Northey also supplied a K-type, with an 8½-inch diameter piston, to Cape Race and according to the Chief Engineer to the Department of Marine and Fisheries:

'… the patentees are now experimenting with a monster instrument having a piston fourteen inches in diameter'.[7]

By 1930, there were 160 diaphones in service in Canada and an increasing number were being installed in the United States. The first diaphone in the United States was not installed until 1914 in New York, unless we include Hope Jones' cathedral organs! Some diaphones were made under licence in the USA, but in the early 1930s the US Lighthouse Service bought the Diaphone Signal Company rights and licensed Deck Brothers, a New York company, to make diaphones for them. A similar arrangement existed between the Canadian lighthouse authority and another company who later made diaphones for them. Before Northey sold the company, some diaphones were made under licence by Steven & Struthers in Glasgow and Chance Brothers near Birmingham.

Diaphone production began in about 1923 at the Chance Brothers Smethwick works, after a few trials with instruments bought from Northey. Within a few months, Chance made their own diaphone.

A single diaphone was installed at Casquets lighthouse in 1921 with a vertical omnidirectional resonator and mushroom reflector and other diaphones followed at Flamborough Head, in 1924 and St Martin's in 1925. Engineering drawings held by Trinity House show a proposed diaphone installation for South Stack dated 1923, but it was not installed at that time. Before 1930, four lightvessels were fitted with diaphones, as were the lighthouses at Hartland Point and the Skerries. Most of these instruments were K-types, the largest size made bar one.

When Northey sold the diaphone rights to the US Lighthouse Service he also sold rights to Chance Brothers for the sole manufacture of the diaphone, except in North America.

Where Chance Brothers were the main contractor for installing a fog signal, they also supplied the engine and compressor sets beside the diaphone, in each case. Every diaphone was made to order and each of the precision instruments varied slightly. Despite the precise machining to specified size and weight, the

---

[6] *Engineering,* journal, 17th September 1909.

[7] F.A. Talbot's 1913 book, *Lighthouses and Lightships* commented that: ' …the builders of this terrible noise producer are experimenting with an apparatus having a piston fourteen inches in diameter'. Despite the assumed experiments the largest diaphone made by Northey's company was the K-type, with its 8½-inch diameter piston.

diaphone casting and the cutting of the slits in each piston and cylinder caused slight variations, which affected the note produced. The castings were gunmetal and machining was to very fine tolerances; the piston fitted the cylinder tightly to within a two thousandth of an inch. The slits were accurately cut in the cylindrical walls of both the piston and cylinder so the piston was essentially a lightweight collection of rings held together by nine ribs.[8]

Each diaphone was supplied with two spare pistons. The diaphone was highly efficient and without mechanical parts; the piston 'floated' in a cushion of air as it oscillated. No lubrication was necessary, although some oil was in any case present in the air from the compressor.

The Canadians were proud of their involvement in the development of the diaphone and the lighthouse authority claimed that after twenty-seven years in service, the first diaphone showed no signs of wear and was expected to last as long again.

There was some wear to the diaphone, however, no doubt much of which occurred during the 'grunt'! With wear, the fine tolerance increased to something quite loose on some instruments. As the piston became loose it was more likely to rattle in the cylinder thus causing more wear; it could even jam, though this kind of damage was usually a result of mistakes in handling. Incorrect alignment when replacing a piston could be costly. I believe that worn pistons were electrolytically treated at Blackwall workshops to deposit a fine build-up of nickel, which was then machined to restore the tolerance. This would have been a difficult job as the lightweight structure, although resilient to the task required of it, was easily damaged.

There were once open fields adjacent to the Chance Brothers factory at Smethwick and when the first diaphone was tested at the works, the local farmer was as upset as his dairy herd by the noise! Each subsequent testing was well notified, but even when kept in the barn the cows did not like it and neither did the farmer! Chance Brothers finally bought the farm for its sports and social club!

Each diaphone made had to be tested and the local press introduced an abusive term to describe the sound which the locals endured during the tests; the '1000 elephant grunt'! When testing of the real monsters took place, windows were shattered by the sound!

Chance Brothers offered the same range of diaphones as Northey, but his larger G and K sizes were additionally followed by the L. No L-types were installed in Britain but they were in service in France, Norway and China. L-types might have been 2000 elephants-worth of grunt!

At the other end of the scale, smaller instruments were made for use at harbours, shipyards, quarries and factories. The smallest size was the A-type, first produced

---

[8]   The smaller sized diaphones had five ribs, the F had nine and the K had thirteen.

by Chance in 1936, but earlier models were advertised in North America as being suitable for military and signalling purposes. The diameter of the piston in the A-type was 1 inch and it weighed 3.3 ounces. By comparison, the K-type piston was 8 ½ inches in diameter and weighed 225 ounces.

All of the smaller diaphones suffered from a design problem which was resolved only with headaches and irritation. A, B and C sizes did not have separate operating and sounding air and therefore did not grunt. A single air supply fed both the oscillating ports and the sounding slits and in this way the smaller diaphone more closely resembled Hope Jones' organ diaphone from 1905. The horizontal operation of the smaller types with only the single air supply, could lead to the pistons stopping in the wrong place and then being held there by the air pressure on subsequent admission, unable to start oscillating. The C-type diaphone used at the Mumbles was notoriously unreliable; even the Admiralty List warned mariners about it!

Despite the difficulties, however, several of the small size diaphones survived for many years and gave reliable service at Wolf Rock, Dover and elsewhere and now that the last two Trinity House diaphones have been permanently discontinued, it is a C-type that can claim to be the last surviving diaphone in use on the British coast.[9]

The single F-type at Portland Bill with the standard resonator distributed sound over 120 degrees of arc. Where a larger sea area was to be protected, the mushroom-topped resonator, as used at Casquets and on the lightvessels, could be used. The mushroom resonator was poorly regarded, particularly by Rayleigh as it did not conform to his theories of sound propagation, yet it proved satisfactory: vertical horns without the reflector on top were always improved when it was added. Experiments with smaller diaphones suggested that without any kind of trumpet they would give a good spread of sound, yet without the resonator the effectiveness was always reduced. The cast iron resonator vibrated at the fundamental frequency and helped to concentrate the sound.

After World War Two, Casquets lighthouse required some rebuilding as it had suffered attack and occupation by both sides. The vertical trumpet from 1921 was replaced by three diaphones and resonators, each giving 120 degrees of sound distribution. An F- and two G-types were installed in an iron turret above the engine room situated in what was originally one of the three light towers. The semi-diesel Blackstone engines at Casquets had survived the war and they remained in service until replaced in 1970.

In the early 1920s Chance Brothers designed fog signal installations using two diaphones to cover a wide sea area. At South Stack and the Skerries, Hartland Point and Dungeness, all installed in the 1920s and 1930s, two diaphones were used with the axes separated by an angle of 100° or 120° and with the mouths of

---

[9] Pen an Glas near Fishguard.

the resonators placed one above the other, displaced by half the wavelength. This design incorporated the physics of Rayleigh to augment the horizontal intensity of the sound by vertically superimposing the fog signal trumpets.[10] As noted in Chapter 8, John Bowen effectively used this theory with the twelve-inch siren by taking two resonators off the same horizontal instrument and spacing them by the half wavelength. As it was a single instrument, there was no problem of synchronization of the output through the horns. With the diaphone installation, however, there were two instruments, one attached to each resonator, and the issue of synchronization was more important.[11]

Although Chance Brothers perfected a means of synchronizing two signals, in the end by delightfully simple means,[12] Trinity House recognised that the asynchronous pairs produced an effective fog signal *because* they were out of synchronization! A rough note was produced but as with the roughness of the 'grunt', this feature seemed to enhance the range and distinctiveness of the signal! As Bowen said

> *'... it (is) impossible to apply known scientific laws rigidly to a specific problem and the choice of fog signal has to be decided largely by empirical deduction and practical experience'.*[13]

The rough note worked. The layout of the installation, type of resonator and choice of diaphone size were all designed to suit the specific location.

The cast iron resonator remained in service for many years but other designs were also used. In the late 1950s fibre glass was used to produce new designs for lightvessels and shore stations. The Humber Conservancy Board lightvessel Number 14, *Spurn*,[14] was fitted with a new omnidirectional design which consisted of two fibre glass saucers placed bottom to bottom with a diaphone situated between them. This design was not widely adopted but it imaginatively solved the problem of producing an omnidirectional diaphone. Instead of the usual design with air entering

---

[10]   Paired cast iron resonators were usually installed in a lozenge-plan iron turret, 13 feet 6 inches by 7 feet, with an overall height of 9 feet 6 inches. At the centre line of the longest side of the turret, the two protruding mouths overlapped, although set at an angle of about one hundred degrees. In the K-type installation, the lower mouth was set on the centre line at 3 feet 3 inches, with the upper mouth at 6 feet 10 inches high.

[11]   Synchronization of twin five-inch sirens was achieved by mechanically linking both instruments with gears or a connecting rod. Without the link, the independently spinning sirens were out of phase and produced beats. Interestingly, the use of beats was considered as a means of enhancing a signal worthy of further investigation by John Tyndall. Beats were again reassessed in connection with the installation of 'air oscillators' in the 1930s.

[12]   Frederick Cooper of Chance Brothers linked the driving air chambers with a section of hosepipe. (Supertyfons were similarly synchronised.) The fluctuating pressures in both casings quickly equalised and regulated the oscillation of the piston. The 'grunt' was tremendous! The synchronized signals, when in phase and spaced at half a wavelength, concentrated the sound level by a factor of three in the horizontal plane.

[13]   Bowen, J.P. *British Lighthouses*, 1947.

[14]   This lightvessel is open to the public at Milford Haven and has a C-type diaphone.

the sides of the cylinder and leaving through the open end, the air was fed into the open end of the cylinder to give a 360° radial flow of sound through the sides and between the saucers.

Another type of glass fibre resonator was the exponential horn with extended vertical axis. Two of these were used at South Bishop, Bull Point, Souter Point and Noss Head in Scotland. This design looked rather like an elastic door which had been sucked in to a small diameter hole in the middle. With the exponential horn, the diaphones were set at equal height and spread outwards at 120° in the turret. The surviving example is the pair at Souter Point lighthouse.

As previously noted, after the war all new lightvessels were supplied with G-type diaphones, but in September 1936, Chance Brothers introduced a variation on the F-type at the Sunk lightvessel. A Trinity House Notice to Mariners announced that 'An experimental Two-tone Diaphone has been installed in place of the station fog signal'. The experimental signal gave a high note followed by a low note, without an interval, twice per minute and Trinity House awaited comments from mariners on the merits of the new signal.[15]

A two-note siren had been used for many years at the Sunk but all of the previous arguments about high and low notes were raised again by the trial two-tone diaphone. John Bowen was unimpressed by the demonstration and the two-tone diaphone was not adopted by Trinity House although one was used at Douglass Head on the Isle of Man. Chance Brothers may have made one or two for export.

In North America, however, two-tone diaphones were widely used and the US Lighthouse Service adopted it as the standard signal. Hundreds were in service between 1930 and the 1960s. The two-tone diaphone was Northey's last development before he sold the company. Northey's son had helped to refine diaphone construction and had also experimented with changing the air supply. The first two-tone was an ordinary F-type, but the refined version had extra oscillating ports which were operated by an auxiliary air valve. Experimental versions used additional, narrower slits and had a longer piston, although the standard F slits could be used. Where the oscillation of the piston normally opened the slits twice with each stroke, the modified piston head only opened the slits once on each stroke, resulting in the usual note of 180 Hz being halved to produce a note of 90 Hz. The two-tone F-type, F2T as it was known in America, sounded quite beautiful, but the musical qualities of the sounds were unnecessary as a distinguishing characteristic and were

---

[15] The characteristic was high note, 1½ sec; low note, 1½ sec; interval, 2 sec; high note, 1½ sec; low note, 1½ sec; silent, 52 sec. The Notice to Mariners went on: 'Mariners are requested to forward the results of their observations of the signal to the Secretary, Trinity House, London E.C.3. These reports should state: (a) How the signal compares with the previous signal, single tone blast which terminated with the grunt (b) Whether or not both high and low notes were always heard (c) If not, which note was heard at a particular time and date, and what was the state of the weather at that time. By Order. J.M. Nicolle, Secretary'

Souter Point diaphones had exponential resonators. The shape of the building reflects the previous Rayleigh trumpets used here previously.

subject to the same problems as the two-note siren. There was always the possibility that one of the notes would not be audible in certain circumstances, possibly with consequences similar to those noted at Lundy with HMS *Montagu*. Although the low notes could sound a little rough at close quarters, without the grunt the diaphone lost its most distinctive quality and for these reasons, the British lighthouse authorities decided against adopting the two-tone diaphone.

When the Chance Brothers company was sold in 1955, diaphone production and other marine engineering moved from Smethwick to Crawley. The Stone Chance company continued to manufacture diaphones and installed signals at Souter Point, Noss Head, Wolf Rock, Strathy Point and Inchcolm. On completion of the new Royal Sovereign lighthouse in 1971, a G-type diaphone made by Stone Chance came into service. The diaphone had an omnidirectional, mushroom reflector and was powered by Gardner 3LW engines and Reavell Quadruplex compressors. When the keepers were taken off in 1994, the diaphone was replaced by an 800 Hz Honeywell electric emitter.

The last diaphones installed at a major lighthouse were at Bull Point on the Devon coast. A landslip in 1972 resulted in the 1950s diaphone installation going over the cliff, along with its exponential resonators. A new lighthouse was completed

Bull Point: the triple diaphone with concrete oval mouths.

in 1974 with a triple F signal. The new lighthouse and fog signal were fully automatic, although keepers lived on the site. A fog detector started up the diaphone when necessary and mains electricity compressed the air, except in an emergency when a Petter diesel drove the Reavell DSA 7L upright compressor.

The Bull Point signal was required to cover a wide sea area and the three diaphones were attached to integral, concrete, oval-shaped resonators, which formed part of the lighthouse tower wall. The oval mouth design returned to Rayleigh's seventy-year-old proposals! The compact interior still contains all of the equipment, except for the fog detector and without too much work this monster could speak again! Instead of the usual green paintwork inside, a pale blue was chosen for Bull Point, making it more like the interior of an NLB lighthouse than a Trinity House building. Bull Point diaphone was discontinued in 1988.

On an April evening, I am sitting on the headland to the east of the lighthouse at Portland Bill. The distant lighthouse is not yet lit. A slight south-westerly wind has brought the fog banks in and as I listen to the gulls and a nearby blackbird, I can anticipate the activity in the lighthouse. Before long, the sound of the first blast of the distant diaphone emerges on my right through the gloom and then the 'grunt' fills the air from all directions before the sound dies away. Surely a fog signal could not sound so wonderful!

*One of the last diaphones installed at a
British lighthouse.*

Chapter 12

Hyskeir

and the last generation of air signals

The Northern Lighthouse Board's Depot at Oban was expanded in the early years of the twentieth century. The stores, workshops and gas-producing plants maintained and serviced many buoys and the lighthouse tenders operated from Oban to service the western coast, particularly the Hebridean lights. Many of the keepers' families lived on the hill above the depot.

For many years, the uncertainties of the boat relief at the isolated lighthouses had been a fact of life but this changed in the early 1970s, with the introduction of regular helicopter links.

On a bright fresh morning in August, whilst the ground staff at Oban refuelled the helicopter and loaded various stores, my colleagues and I donned the orange survival suits and watched the safety information video. The video was shown before each flight and its repetition guaranteed a thorough familiarity with its contents such as which parts of the helicopter it was unsafe to be near or climb on, how the doors operated and how to prepare for impact should the helicopter ditch in the sea, at which point the survival suit would come into its own!

Guests enjoying breakfast in the hotel behind the depot looked on as the last items were packed inside the little red helicopter. There was a strong smell of kerosene as we climbed aboard the Bolkow 105.[1]

Once all the instruments were checked, the engines were started and the rotors began spinning. The communications headset helped to reduce the noise level but it remained noisy and became worse as the rotors increased speed, dipped forward and the helicopter lifted. After hovering for a few moments, the helicopter swung around and over the Sound towards the island of Kerrera, above the buoys and boats in Oban bay. Cruising at 130 mph, the flight to Hyskeir would take thirty-five

---

[1]   Bond Helicopters are chartered by the Northern Lighthouse Board and Trinity House.

minutes, mostly over the sea, first crossing the Firth of Lorne towards Mull and Lismore and then following the Sound of Mull before heading towards the Ardnamurchan peninsula.

Hyskeir is 23 miles north-west of Ardnamurchan lighthouse. As we cleared the coastline our pilot thought we might see some whales, but we were unlucky, although we did get an impressive view of huge basking sharks near the surface of the dark water. Hyskeir came into view and shortly afterwards the helicopter flew over the rocks and grass and put down on the concrete landing pad. The helipad was built beside the tower at Hyskeir, as at all of the Scottish rock lighthouses. Even at the isolated Dubh Artach and Skerryvore lights there was just sufficient space for a landing pad on the rocks.

At most of the Trinity House rock stations, there was insufficient space beside the lighthouses or they were too frequently wave-washed, and engineers constructed the landing pads on the towers above the lantern. The Smalls lighthouse originally had a landing pad built on the rock in 1972, but subsequently joined the seven other Trinity House lighthouses fitted with helidecks on top of the tower.[2]

The small diameter of the helideck is only appreciated when standing on it! Wolf Rock was the first lighthouse to be fitted with a helideck in 1973. Before the introduction of the helicopter reliefs, keepers at Wolf Rock were often trapped when the tender was unable to approach in violent seas – a stay of over two months was not uncommon. When the boat relief *was* possible at the rock lights, the keepers and all other items were winched from the boat on to the landing stage of the lighthouse. This was a hazardous procedure for the keepers and an unsuitable one for transferring sensitive equipment – everything had a hard time getting on to the landing and for the keepers it usually meant a soaking.

Prior to the regular helicopter service to Hyskeir, a landing stage on the south-west of the main island could be approached by boat in fair weather.

The Oigh Sgeir, which means 'high rock', barely stands thirty feet above mean high water but is large enough to accommodate the lighthouse, engine room and keepers' quarters on the most elevated part of the land. The island provided sufficient additional space for the helipad and vegetable gardens. A very small 'golf course' was purportedly maintained but a hefty drive from the tee would probably have lost the ball in nearly every direction! On my visit, the garden was full of potatoes, carrots, cabbages and beans.

The rocks at Hyskeir form five separate islands and the three largest have equal portions of grass and bare rock. These are linked by bridges at the narrowest and most sheltered spots, and at low water these three virtually form one island. The other two have deep water surrounding them. The rocks were formed by volcanic

---

[2]   Bishop Rock, Longships, Wolf Rock, Eddystone, Needles, Hanois, Casquets and Smalls.

activity stretching from Mull and Rhum over the Hebrides as far south as Northern Ireland. The Giant's Causeway in Antrim has echoes on Hyskeir, where the exposed rocks show the same form of hexagonal columnar jointings, though less spectacularly. Near the extremity of the grass cover, the dark rocks are covered in thick orange and sage-coloured lichens.

On a good day, between the islands of Rhum and Canna, the cloudy peaks of the Cuillin Hills on Skye are visible over twenty miles away. The island of Rhum is visible ten miles to the east of Hyskeir and to the north-east is the island of Canna, five miles away. A submarine ridge extends towards Hyskeir from Canna, but the depth of the surrounding sea floor is between forty and sixty fathoms. At Hyskeir's edge the water is nearly thirty fathoms deep and a few miles south-west, towards the Atlantic, the Hebridean sea floor drops to one hundred fathoms. Looking at the dark sea I sensed the vulnerability of the low-lying land: although calm, the vast sea still appeared to threaten.

The sea was almost noiseless, but the relative peace and quiet was rich with the wonderful trilling of oyster catchers, terns and plovers. There were herring gulls and black backed gulls inshore and, just off the land, guillemots and gannets. The terns swooped noisily and aggressively at times. In the receding tide seals lounged on the rocks until alerted to my presence by the terns, when they lumbered and splashed into the water. A few heads bobbed in calm water and watched and snorted. The other sound was quite different, a regular medium pitched note which reverberated in the air and disappeared towards Canna without an echo every thirty seconds.

The supertyfon, installed at Hyskeir in 1964, was preceded by an acetylene fog gun, in service from 1937. The Northern Lights used acetylene guns at a number of stations, with about fifteen in service from 1937. One was installed at Out Skerries as late as 1946. The early automatic operation of many of the Scottish minor lights with acetylene was supplemented by the gas gun fog signal. Acetylene gas could be produced on site by immersing calcium carbide in water, although this kind of plant was phased out after World War Two in favour of bottled gas delivered by tender.

The Irish lighthouse engineer John Wigham devised an explosive signal designed to use town gas, which he offered to Trinity House for their evaluation in the early 1880s, but the device was not adopted in England.[3]

The acetylene gas gun was manufactured in several sizes by Moyes of Glasgow. The range of the signal was less favourable than the tonite explosive charge, but it was cheaper to run and could be fired with greater frequency.[4] Acetylene mixed with air in a detonating chamber was ignited by an electric spark and exploded outwards through a conical resonator. A spark could be produced by a battery or by a flint on a mechanical apparatus. The gun was used on some acetylene buoys and

---

[3]   Wigham's gas gun was demonstrated in 1882 and is described in April 7th edition of *Engineering*.

[4]   Figures quoted to the 1929 Lighthouse Conference suggested that if the smallest gun was to be fired every minute for 365 days the costs would have been about £20 and if the largest gun were used this would have cost £150.

radio-controlled acetylene guns were in use on the Clyde in the 1930s. A Moyes gun was installed at Point Lynas lighthouse on Anglesey in 1948.

The acetylene gun at Hyskeir was discontinued in September 1964 when the supertyfon came into service. The new horn house was built on the north tip of Hyskeir, less than half a mile from the lighthouse. Unlike the large bore iron pipework which linked earlier Scottish lighthouses to their compressed air fog signals, a small bore flexible pipe connected the horn house with the compressors in the engine house. Although the horn house was built on the highest ground, its exposed site could be lashed by the sea and in 1968 the signal was put out of action by storm damage for several months.

Hyskeir lighthouse was converted from oil to electricity in 1959, the same year that repairs to the fire damaged Skerryvore lighthouse were completed. The supertyfon fog signal, which was installed at Skerryvore at that time, was the first of its kind to be used at a Scottish lighthouse, although similar equipment had recently been installed at Eddystone lighthouse.

Hyskeir lighthouse was 'Erected by the Commissioners of Northern Lights' in 1902 to the designs of 'D.A. Stevenson, Engineer', as the carved lintel above the original entrance to the tower proclaims. This doorway is now in a lobby off the engine house. Inside the engine house there are three Gardner 3LW diesel generator sets and the fog signal compressors. The engine exhaust could be faintly heard at the top of the tower, along with the quiet hum of the engine room from inside the building. In the wooden-panelled service room beneath the optic, the quiet ticking of the brass clock was louder than the engines and the birds and the distant supertyfon.

If the diesels all failed, the electric lighthouse could fall back on the emergency gas lamp – the clockwork machinery below the optic was still operational on standby. The weights were suspended in open chains through circular ports in the floors of the upper rooms of the tower.

The massive hyper-radial optic was made by Chance Brothers to Stevenson's designs and almost completely fills the lantern. Where the reflecting prisms would normally produce a convex curve above the refracting prisms, here, the panels of reflecting prisms are eccentrically inclined by 11½° in a concave configuration. The elements of this optic were probably being assembled in Smethwick alongside those produced for Portland Bill. The optic at Hyskeir was installed in 1904.

Back in the engine room, opposite the Gardner generators, were the electric motors attached to the two Hydrovane compressors. The compressors ran at 1500 rpm and could deliver air at up to 100 psi. When the last Scottish siren was discontinued, the remaining air signals in Scotland all had this kind of turbine compressor. This design was also installed to replace older compressors at some Trinity House lighthouses.[5]

[5] Holman and Worthington Simpson compressors were also used in the Trinity House supertyfon installations, usually driven by Brook or Crompton motors.

A small, steel air receiver in the engine room was open to the pipe which lead to the horn house, where there were two similar receivers. The fog signal was regulated by an air-operated timing device and the flow of compressed air was capable of producing an accurate interval between blasts. This had been the means of regulating the blasts in the early sirens, and the elaborate system installed by Charles Ingray at Ailsa Craig is worth remembering. At Hyskeir, pressure in the main air pipe sealed the flow of air until pressure built up in a small counterbalancing tank, to the point where the force operated a small piston which tripped open the counterpressure valve to the supertyfons.

Several lighthouse authorities tested diaphragm horns in the 1930s after different manufacturers introduced the instruments. The sound is produced when a metal diaphragm is rapidly vibrated by electromagnetism or compressed air. The Swedish firm Kockums developed the tyfon air horn from the 1920s and it was increasingly used aboard shipping, as diesel engines replaced steam turbines and air signals replaced the steam whistles. Some other air horns developed for shipping[6] also found their way into service at lighthouses and harbour lights later on, such as the Secomak signals[7] and Airchimes.

The supertyfon signal at Hyskeir comprised two TA 150/255 horns mounted above the horn house on a metal framework. With each blast the compressed air escaping through the horns rapidly expanded, briefly lowering the surrounding air below its dew point to form little clouds of condensation.

The mouths of the horns were horizontally separated by about 120° and the diameter of the mouth was 13½ inches. The horns were 18½ inches long. The design of this exponential resonator stabilised the frequency of the vibrations of the diaphragm by regulating the escaping air pressure waves.

There were two chambers inside the casing of the supertyfon. Compressed air supplied to both chambers held the diaphragm (a double one) sealed against the throat of the resonator. When the counterpressure in the rear of the horn was momentarily dropped, the higher pressure in the front chamber impinged on the

---

[6] The length of a ship determined the frequency and range required of the ship's whistle* and in 1972, international regulations agreed that vessels over 200 metres should sound a note between 70 and 200 Hz, having a range of at least two miles; for vessels over 75 metres, 130 to 350 Hz, range 1½ miles; for vessels over 20 metres, 250 to 700 Hz, with a range of one mile and for vessels under 20 metres, again, to sound between 250 and 700 Hz with a half-mile range.

 * The term 'ship's whistle' dates from the time when a steam whistle was the usual instrument. At some stage the distinction between fog horn and ship's whistle has blurred. Without steam-driven ships the 'whistle' was usually an air horn.

[7] Secomak produced 'whistles', sirens and electric piston horns. Still in use at Cardiff Docks as the barrage project began, a Secomak piston horn signal stood at the Alexandria Dock entrance. Its low note was resemblant of a motor-driven siren or even a diaphone, as it was referred to in the List of Lights for many years. Secomak sirens were used at one or two Scottish lights, including Crammag on the Mull of Galloway and Davaar in Campbeltown Loch; similar equipment made by the Klaxon company was trialled elsewhere.

Hyskeir's supertyfon horns.

diaphragm pushing it off the seating and allowing air to escape through the horn. Fluctuating pressure variations on both sides of the diaphragm then caused it to vibrate, acting as a valve, rapidly opening and closing the horn throat to release bursts of air and produce the sound waves in the horn. The air-operated timer in the horn house supplied the counterpressure air line, and the sounding air to the front chamber was supplied directly from the receivers.

Modern compressors and compact equipment capable of operating at high pressures enabled air signals to be reconsidered as fog signals in the confined spaces of rock lighthouses.

After the supertyfon was installed at Skerryvore, the Northern Lighthouse Board installed another at Cantick Head in 1962 before Hyskeir and Bell Rock in 1964. After Eddystone and the Hanois, Trinity House installed supertyfon signals at two mainland sites, St Catherine's, 1962, and Trevose Head, 1963. At both of these sites other air signals had been in use – a diaphone at St Catherine's and Rayleigh's experimental trumpet at Trevose – and the existing air receivers were retained, although new engines and compressors were installed.[8] Eight supertyfon horns were installed in the iron turrets at both lighthouses in two banks of four. At St Catherine's two complementary sizes of horns were used sounding at 205 and 260 Hz.

---

[8]   The familiar Quadruplex compressor was not used with any of the supertyfon signals and at Trevose Head, Reavell DSA7 compressors were used, which had two large bore vertical cylinders of different sizes. Running at 525 rpm the double acting pistons drew air into the larger cylinder and the compressed air was then let into the smaller cylinder to compress the air further. The heat generated by compression could be up to 300° Fahrenheit in the cylinder head. The compressor delivered air at 30 psi but was capable of higher pressure. Valves on the storage receivers were connected by small bore copper pipe to the compressor and if the pressure in the receiver exceeded the specified limit the compressor was automatically unloaded. If dangerously high pressures built up, the system fail-safe actuated a solenoid switch to shut off the power supply to the motor, or the diesel to the engine.

One of the supertyfon exponential horns used at The Needles.

At Lundy South, eight two-note horns were installed to replace the explosive signal in 1964. The two-note supertyfons were not sounded separately, as the two-note siren and the experimental two-tone diaphone had been, but instead sounded simultaneously to produce a chord. The manufacturers claimed that strong harmonic overtones were produced by the supertyfon, even if only a single horn was used.

At the Needles, two banks of two single-note horns were installed to replace the reed horn. The air receivers from 1946, installed below the lantern room, remained in service as did the single cylinder Gardners and the rheostat, through which the current supplied the compressors. The replacement Hydrovane compressors were in use until the signal was dismantled in 1994.[9]

The automatic operation of the supertyfon at Hyskeir was effectively controlled by compressed air, but Trinity House, having converted their major lights to electricity, used the electric coders they had developed in the 1960s. Where the coder ordinarily tripped a solenoid valve to open the admission valve to the sirens and diaphones, with the supertyfon an electric signal from the coder to a specialized

---

[9]   Needles keepers were taken off in December 1994 and the supertyfon was subsequently removed to the Lighthouse Museum in Penzance. The supertyfon was replaced by Honeywell ELG 500 electric emitters.

Air connections to two of the eight trumpets at Trevose.

actuating valve was required. This valve controlled the counterpressure air to the rear of the tyfon. A spring tension screw on the valve enabled accurate adjustment of the pressure for optimum working.

The Trevose supertyfon had eight horns and the two banks of four had one actuating valve each. All four emitters in each bank were additionally connected by equal lengths of sealed piping, to maintain synchronization between the fluctuations in pressure and the resulting sound produced in each horn.

To protect the supertyfon from dirt and rust particles, besides the usual oil and water suspended in compressed air, Norgren air filters were usually installed in the air main near the diaphragms. Some of the later high speed compressors were designed to use a lot of oil in normal operation.

Small high pressure steel receivers were installed with most of the new supertyfons and on the rock lights the receivers were fitted in the upper floors, attached to the underside of the floor of the lantern room or the service room. At Eddystone, the receivers were 6 feet 9 inches long with a diameter of 1 foot 3 inches and one was placed near each of the sounding units. At Bishop Rock there were six receivers in the service room, each with a capacity of 8.2 cubic feet at 100 psi. The capacity of one of the large riveted iron receivers at Trevose Head, by comparison, was 265 cubic feet at 35 psi and was originally designed to safely exceed a pressure of 150 psi.

Two other Scottish lighthouses were equipped with supertyfons in 1966, namely, North Ronaldsay and Ailsa Craig, where the north and south horn houses were finally abandoned when the new horn was located on the lighthouse tower.

Trinity House installed supertyfons to replace the explosive fog signals at Longships in 1969, and Smalls in 1971, but it was not until 1976 that Bishop Rock was fitted with a supertyfon when the explosive jib was unlovingly hurled over the side! On the automation of the rock lighthouses, the remaining supertyfons were replaced by electric emitters.

The planned life expectancy of automatic electric equipment is fifteen years. When Trevose and Hyskeir were automated they shared the record for supertyfon longevity in service at nearly thirty-four years.

The Northern Lighthouse Board installed another kind of diaphragm air horn at four sites including the Mull of Kintyre, between 1968 and 1976. 'Air chimes' were originally made in Canada[10] and were widely used on ships – the *Ark Royal* amongst many others had a combination of the largest types of this instrument. The diaphragm was vibrated in a similar way to the tyfon but with only one air supply into the emitter. On ships and at lighthouses, the horns could be combined not only to increase the power and range of the signal, but also to produce distinctive 'chimes'.[11] Six types were available from the lowest note horn, 85 Hz, to the highest note, 250 Hz and the combination of notes could be very effective. At Stroma, twelve Air Chime horns sounded two notes and at Fair Isle nine horns were installed. The Air Chimes could be operated at pressures varying from about 50 psi up to 150 psi, intended to give the signal increased range.[12]

The NLB siren sounded at about 120 Hz but slightly higher notes were chosen for the Air Chimes[13] and tyfons. At the Mull of Kintyre, three Air Chimes were installed in the distant horn house in 1976 and, as with the supertyfon at Hyskeir, the blasts were regulated by air controlled timing valves.

Air Chimes were tested in the summer of 1967 at Dungeness when it was still the main site for Trinity House experiments. The effectiveness of the two largest sized horns, sounding at 110 Hz and 85 Hz,[14] were evaluated but, regardless of the success of these trials, by this time, Trinity House was already a long way down the road to equipping lighthouses with the powerful electric fog signal on which they had spent some time and money: the triple frequency emitter stack.

---

[10]   In the UK air chimes were made under licence by Smith Dennis, a company owned by relatives of Sydney Smith of Nottingham, who manufactured sirens and valves at the end of thr 19th century.

[11]   The manufacturer's promotional leaflets said: 'When combined as chime toned whistles, their melodious sound is comparable to none. They command attention and add prestige to all ships and plants wherever they are used'!

[12]   Mull of Kintyre sounded at 60 psi, Fair Isle sounded at 80 psi.

[13]   The supertyfon sounded a fundamental note between 150 and 260 Hz.

[14]   Air chime dimensions: 85 Hz signal, 67 inches overall, diameter of mouth 24 ins. The 110 Hz signal: 53¾ ins overall, mouth diameter 22 ins. The stainless steel diaphragm was 7½ ins in diameter for all horns – the difference in the notes was produced by the dimensions of the horn, as with the supertyfon.

Chapter 13

South Stack

and the electric fog signals

The pros and cons of electric fog signals were discussed at the first International Conference of Lighthouse Authorities in 1929. Despite the development of the radio beacon and the perennial question of the reliability of sounds transmitted in air, sound signals continued to provide an essential service. Powerful, low frequency, compressed air sirens and diaphones were used at major sites to produce the long range signals, although some short range electric signals were in limited use at harbour approaches between the wars. Within the next generation, technical developments and changes in the required performance of fog signals resulted in electric 'nautophones' replacing some compressed air signals, even at major sites.

The AGA company produced the one kW 300 Hz electric signal in the 1950s in which two steel diaphragms were vibrated by alternating current at 150 Hz. The LIE 300 unit had an iron core surrounded by coil windings, and the electromagnetic oscillation of the two attached diaphragms generated air pressure waves in the resonating horns attached above and below. The aluminium or iron unit was available with omnidirectional or directional horns and could be vertically stacked to increase the power and range of the signal up to four miles.[1] Each emitter unit required a separate drive unit. The CEFA 1000 drive unit regulated the voltage supply and monitored the one kW output from the emitter. The unit also had an integral coder. In multiple emitter stacks, the drive units used master and slave arrangements to prevent the emitters from sounding out of phase and producing 'beats' or partial cancellations.

---

[1] In the 1960s the International Association of Lighthouse Authorities (IALA) determined the 'usual range' of a fog signal. This was defined as the distance in calm, foggy weather at which a lookout outside the bridge of a ship would stand a 50% chance of hearing the signal above the shipboard noise level found in 50% of large merchant vessels. Trinity House subsequently defined three categories of fog signals based on the IALA 'usual range'. 'Category A' signals have an IALA usual range of 3 miles; 'Category B' – IALA usual range of 2 miles; 'Category C' IALA usual range of one mile. The 4 kW LIE 300 signals had a minimum range of 3½ miles.

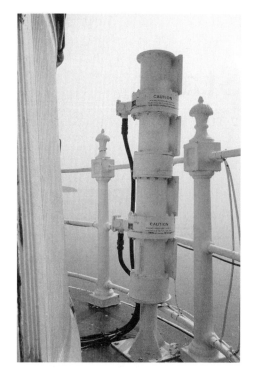

Double unit ELG 300 emitter, one of two stacks at Flatholm lighthouse.

This fog signal was very successful and was in use for over twenty-five years at some stations. The same basic signal was still being installed by AGA's successor, Pharos Marine, in the 1980s. The LIE 300 was in service at over twenty British harbours and major lighthouses, from Cape Wrath to Hartland Point, and it replaced the supertyfons at Skerryvore and Eddystone.

The so-called 'pure tone' electric emitters produced a sound which lacked the rich harmonics[2] of the compressed air signals. Critics recognised that harmonics improved the audible range of a fog signal and assisted in distinguishing the sound from background noises.

By the 1950s developments in the use of radar and the continuous transmission of radio beacons in clear as well as foggy weather indicated inevitable changes regarding fog signals, even though it remained the policy of lighthouse authorities to provide powerful sound signals. It was recognized that the low notes of the siren

---

[2] Harmonics are the multiples of the fundamental frequency sounded. For example, a siren sounding at 200 Hz would produce harmonics at 400 Hz and 800 Hz and 1600 Hz, etc., as well as 100 Hz and 50 Hz, etc. The combination of the harmonics with the fundamental note produce complex wave forms to give a particular character. The 'pure tone' signals produced harmonics but they were much less pronounced.

and diaphone could be lost in shipboard noise levels and tests at sea established that sounds between 300 Hz and 800 Hz, were more likely to be heard.[3]

In recognition of the above considerations, research at Trinity House sought to develop a new type of fog signal that would use tuned electric emitters to sound simultaneously at three frequencies between 300 and 800 Hz. Experiments began with sound emitters as used in public address systems purchased from the Tannoy Company. Tannoy subsequently modified their pressure units to provide custom-made 100 W units with a standard metal resonator and these units were combined in vertical stacks.

In 1954, thirty-five 20 W emitters were incorporated into a trial fog signal established at St Anthony Head lighthouse near Falmouth. Three fundamental notes were simultaneously sounded by the new fog horn. A similar version was constructed at Dungeness experimental station where the equipment consisted of eighteen 100 W emitters in three columns of six. The frequencies of the three notes sounded were 390 Hz, 420 Hz and 630 Hz.

In 1956 a new signal was built at Dungeness with the emitters installed in specially designed, cast, concrete blocks. Each block was two feet square and one foot thick with the mouth of the emitter in the centre of the cast exponential horn shape. The concrete horns were designed to be used as building blocks and when assembled, the horn mouths were spaced according to the established principle of separation by half the wavelength of the note sounded to augment the output.[4] The signal at Dungeness had thirty blocks in five columns of six and the columns formed a curved wall to give a wide spread of sound. The new signal was compared with a diaphone, siren and supertyfons in July 1958 in trials not unlike the previous ones made at Dungeness over the years. During the course of the trial period, observations made aboard ships favourably indicated that the selected frequencies of the new signal were distinctly audible above shipboard noise. The design was approved for service use at Dungeness and similar equipment was approved for installation at North Stack fog signal station at the same time.

Production of the fog signal was developed further in collaboration with Stone Chance and a thirty speaker signal was in service at North Stack in 1958. In the following year, a third signal was constructed at Lynmouth, followed by the new lighthouse under construction at Dungeness. As noted in Chapter 2, the new lighthouse, which opened in 1960, had sixty speakers built into the wall of the tower just below the lantern. This signal replaced the prototype thirty-speaker unit.

[3] Tests had been carried out by many lighthouse authorities and, as early as the 1930s, research by the Atlas and Telefunken companies, in co-operation-with the German lighthouse service, established that medium notes were more usually heard aboard ship than the low notes. Electric emitters sounding at 150, 250, 500 and 1000 Hz were used in these tests.

[4] Whether the sound was augmented by this theory or not is unclear as the notes between 400 and 750 Hz, the designed range, have wavelengths less than the horn spacing, between 2 feet 10 inches and 1 foot 6 inches.

The triple frequency electric fog horn installed at North Stack in 1958.

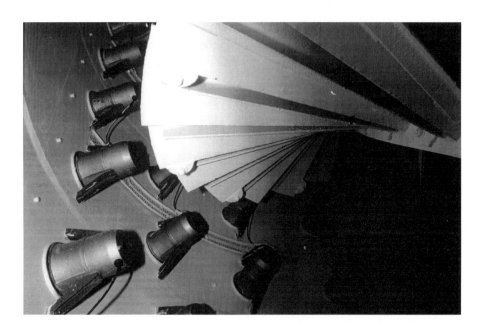

The staircase and rear of the 60 fog signal speakers in the tower at Dungeness. (See photograph of tower on page 16.)

Between 1958 and 1971, Trinity House installed a total of eleven triple-frequency fog signals. The frequencies of the selected notes varied with different signals but they were generally around 400, 550 and 700 Hz and the sound produced was more or less discordant at different stations. The chord produced had an arresting nature which Charles Babbage had previously anticipated: '... Some combination of discordant sounds may be most effective ... because contrasts produce stronger impressions ...' Babbage also wondered about sounds which '... set the teeth on edge' and perhaps '... highly obnoxious sound' would penetrate further and be more easily noticed! This might be a description of the triple-frequency fog horn except for the date of the article, 1852!

The three notes were produced by stimulating the Tannoy pressure units with current from separate windings on the stator of an alternator, so designed as to enable power to be simultaneously taken off at three frequencies. The alternator operated at 3000 rpm and was driven directly by an AC motor at those lighthouses on mains electricity, or by a diesel engine at others. Although the alternators suffered heavy wear, this type of fog signal was effective over a wide sea area and had an impressive range which exceeded other electric signals and compared reasonably with the compressed air signals.

Trinity House engineers developed a range of electrical monitoring equipment for lighthouse machinery in the 1950s and 1960s, including coding and transmission equipment and an automatic fog detector. An increasing number of lighthouses were connected to the National Grid, 240V or 415V, but also had back-up diesel generators. Some electric lighthouses operated only on power generated on site, having no connection to the mains. At both kinds of electric lighthouses, batteries were on charge for back-up and for use with low consumption electrical equipment. With periodic modernisation, the cumulative result was that many lighthouses had a variety of equipment designed to operate on 100V DC and 240V AC. Excepting the early electric lighthouses established in the nineteenth century, the electrification programme began in the 1920s. South Stack lighthouse was electrified in 1938.

The fog detector was designed to monitor visibility and automatically start a fog signal when necessary. Its early implementation caused some difficulties at sites without electricity as the choice was then to run generators or convert the new machine to battery operation. In the early 1960s, the electrification of five lighthouses, which were to be remotely monitored from Holyhead Depot, was proposed. Each of the lighthouses was to be equipped with a triple-frequency fog signal attached to an automatic fog detector.[5]

The fog detector was developed in the 1950s on the principle that water droplets suspended in foggy air would scatter light. The back-scattered light from a narrow beam was detected on a photoelectric cell at the back of a telescope and a signal

---

[5] Holyhead did ultimately monitor a number of automated lights and triple-frequency signals were installed at North Stack, South Stack, Bardsey and St Bees.

proportional to the intensity of the back-scattered light was amplified and calibrated. When a predetermined level was exceeded, the fog detector would trigger a relay circuit and start the fog signal.[6] The fog detector, which stood about three feet high, slowly scanned the horizon between any pre-set points and also made periodic checks of its own operation. The design was subsequently produced by Stone Chance under licence from Trinity House. Reliable operation of the fog detector was essential and a cautious Trinity House made provision to check each new piece of equipment at Dungeness prior to its installation at the appointed lighthouse.

The new lighthouse at Tater Du in Cornwall was opened in 1965 and, like Dungeness, the design incorporated the fog signal speakers in the wall of the tower. Tater Du had seventy-two speakers and was built at about the same time as two other triple-frequency signals, one on the island of Bardsey and the other at South Stack. Although Tater Du was connected to mains electricity and fully automatic, the alternators for the fog signal were not driven by electric motors, as at Dungeness, but by diesel engines. Where the electrically-driven alternators could be started quite quickly and soon got up to speed, the diesel start was slower and the two cylinder Ruston engine also ran more slowly, at about 1600 rpm. To bring the alternator up to the necessary speed of 3000 rpm, it was coupled to the diesel engine via gearing and an hydraulic clutch. Lynmouth had the first alternator driven by diesel engine and the others were at Bardsey, 1965, Lundy, 1969 and St Ann's, 1971. These were all manned lighthouses but Tater Du was remotely monitored.

The engines at Tater Du and Bardsey had large heavy flywheels which were designed to maintain the engine speed over the load of the alternators – the load on the engines produced an audible slowing of the rotor during each blast. The heavy flywheels achieved their objective but placed a great strain on the engine when starting up. Over time the torque stresses on the shaft caused cracking and at Tater Du, there were dramatic consequences when the shaft broke. Hydraulic clutches were subsequently introduced. Similar two cylinder Ruston engines also powered the alternator at St Ann's lighthouse from 1972, when the triple-frequency signal replaced the last two-note siren which had been in service there for nearly seventy-five years. The new triple-frequency signal at St Ann's was the last one to be introduced into service and is presently the only survivor, excepting Dungeness. St Ann's produced 6 kW of sound energy.

Once there were plans for a more powerful version of the triple-frequency signal which would have used specially made electro dynamic emitters of 500 watts rather

---

[6]   The back-scatter principle is still used in fog detectors and from the mid 1970s the FD 300 became the most widely used fog detector at British lighthouses, including South Stack. The FD 300 compared calibrated readings between a pilot light and light from the main infrared emitter reflected by back scatter. The smaller FD 310 succeeded the FD 300, but also used back-scattered infrared light sampling for 12 seconds every 2 minutes. This equipment can be operated on lower current, such as from battery or solar panels. The FD 300 and the FD 310 were made by Pharos Marine. New compact designs installed at a number of lighthouses also measure scattered light but with only a narrow angle of deflection.

than the standard 100 watt units, but these plans went no further than the prototype stage.

At South Stack lighthouse, the seventy-two emitter signal was built in front of the tower, about 30 feet from the site of the earlier fog signal building. The previous signal, a diaphone, had been connected by underground pipe to the engine room in the main building at the foot of the lighthouse tower. The compact electrical equipment for the new signal occupied much less space than the previous Blackstone engine and compressors. There was a small cabinet for the coder and there were two motor and alternator sets accompanied by the main control cabinet. A duplicate control cabinet stood in the Trinity House Depot at Holyhead, an echo of the proposal to remotely monitor five stations from the depot.

The control panel monitored current supplied and drawn and incorporated alarm circuits to check emitter failure and current fluctuation. The alternators at South Stack were made by Mackie of Lambeth, who were involved in the development of the Trinity House signal from an early stage. Lancashire motors, supplied by three phase alternating current at 415 volts, produced 15 bhp to drive the alternators directly at 3000 rpm.[7]

Despite the intensity of the sound level near the fog signal stack, inside the building itself it was relatively quiet. Each emitter was supplied with current via a control cabinet containing orange neon lights, which represented each emitter and clearly indicated a faulty circuit or speaker. The new fog signal came into service at South Stack on October 1st 1964.

One and a half miles north-east of South Stack lighthouse stands North Stack fog signal station. Before the electric signal was installed at North Stack in 1958, a fog signal gun had been in service there for over a century. The battery was first established in the 1850s and was taken over by Trinity House from the Admiralty in 1857, when two eighteen-pound guns were in service, similar to those at Flamborough Head and Lundy. The gun was '... fired every fifteen minutes when packets due, otherwise every half an hour'.[8] The 1865 Admiralty List included in the reference to South Stack that 'A gun is also fired from the North Stack every hour and half hour during foggy weather; and when the Mail Packets are expected every quarter of an hour'.[9]

Approaching Holyhead in fog, the packets would hear another fog signal warning from the lighthouse on the end of the long Breakwater. This lighthouse had a fog bell which was supplemented from the late 1870s by rockets fired in reply to the North Stack gun when steamers were expected.[10] Rockets were also tested at North

---

[7]   Smaller 30-speaker signals used 7.5 bhp Brook electric motors.

[8]   House of Commons *Review of Lighthouses*, 1873.

[9]   The Packet Service was established in 1689 and carried mail and official documents all over the world. The packet service into Holyhead primarily communicated with Ireland.

[10]   The 1½ mile long breakwater afforded protection to mail packets and paddle steamers which linked Ireland to London via the North Western Railway Company's terminus at Holyhead.

Stack[11] but, unlike at Flamborough Head and Lundy, here the gun was retained. A supplemental explosive fog signal was introduced, however, and fired alternately with the gun. The muzzle-loaded guns were replaced in 1920 by a breech-loading gun capable of more frequent firing,[12] but it was, nevertheless, only fired once in ten minutes and once every alternate ten minutes the explosive charge was detonated.[13]

Another gun had been installed at North Stack in 1876, the so called 'special gun'. After the South Foreland fog signal experiments in 1874, Professor Tyndall had been obliged to proceed with gun trials. Guncotton and rockets were incorporated into this programme. Tests consistently proved that a 5½-inch Howitzer produced a superior report to the 18 lb cannon with the same 3 lb charge. Tyndall thought that the shape of the gun might be further modified to improve the effectiveness of the sound and James Douglass believed that

> '... the addition of a conical iron projector or trumpet similar in form to (that of the) Syren now at the South Foreland, would increase considerably the efficiency of the gun'.[14]

His comments were based on test firings through the detached siren trumpet at South Foreland. Douglass, like Tyndall, was convinced of the superiority of the siren but was not unwilling to experiment with gun signals. Traditional supporters of the guns, including influential Elder Brethren members, looked favourably on suggestions to develop a specially designed fog gun. Correspondence between Trinity House, the Board of Trade and Woolwich Arsenal followed and four guns were developed for test firings.

A parabolic muzzle was designed and Tyndall thought this would augment the sound of the gun in two ways, firstly as a trumpet in directing and concentrating the sound, and secondly by vibrating, rather like a bell. (Tyndall also believed that the benefits of the second property would not be detectable except at relatively close quarters!) Major Maitland of the Royal Gun Factories designed scale models and then full scale versions for the tests at Woolwich. Tyndall suggested the use of bronze for its improved 'ringing' quality, and the four guns tested were: bronze parabolic, cast iron parabolic, bronze conical and cast iron with normal muzzle. In trials, the guns were fired at five minute intervals with the observers moved further

---

[11] Out at sea, in February 1877, rockets were compared with the gun and the Skerries siren.

[12] In 1984, the gun was recovered from the sea by divers and found to have been made in the 1890s. It was traced back to previous service on two ships prior to installation at North Stack in 1920. The gun had been despatched over the cliff when the electric fog signal was installed. See *Lamp* (Newsletter of the Association of Lighthouse Keepers), No. 29, December 1995.

[13] The alternate firing of the gun and explosive signal every five minutes is recorded in the Admiralty List from 1929 to 1953.

[14] Miscellaneous correspondence and reports, 1874-1878. Unpublished Trinity House documents.

The special gun used at North Stack

away each time recording the relative loudness of each firing.[15] Maitland went on to design and construct a breech loading gun with a parabolic muzzle.

The 'special gun' had an overall length of 65 inches, with a parabolic mouth 48 inches in diameter and 14 inches deep. A 'breech-piece' with five chambers[16] revolved horizontally into the breech loading position and into which the cartridges were loaded. The bore was five inches and the chamber seven inches deep. This gun was much easier to work than the muzzle-loaded cannon and after some trial firings, it was determined that this extraordinary device should be transported to North Stack for a trial period. By December 1876, Maitland submitted his report to Trinity House on the success of the gun, but the success was sadly limited. The gun only fired a few rounds at North Stack before serious cracks appeared in the metal. The gun had previously fired a total of 500 rounds and may have sustained damage in transit to North Stack over Holyhead mountain. Repairs obviously proved unsatisfactory since only a further 50 rounds were fired before the cracks made the gun too dangerous to be used. After the special gun was discontinued at North Stack there was no further attempt to modify it or replace it and the 18 lb guns were restored to service.

Two gunners and their families lived at the fog signal station and the battery still stands on the lower cliffs in view of South Stack. The magazine has the Trinity House crest over the door with the date 1861. Nearby is the electric fog signal which succeeded the battery and remained in use until the 1980s.

---

[15]   The guns were fired in an unknown sequence and the observers were instructed to assign '5' to the first report and submit higher or lower relative values to successive firings.

[16]   E. Price Edwards, in his 1884 book, *Our Seamarks* described the gun as having six chambers, but drawings at Trinity House show five. The special gun was unique so it is something of a mystery that *The Engineer* (1st January, 1875) described the parabolic mouthed gun with a '... breech-piece revolving vertically behind the gun', rather than horizontally. This article did not specify the number of chambers.

The magazine at North Stack fog gun battery.

In 1907, a submarine bell was installed 1500 yards west of North Stack which gave 5 strokes every 20 seconds, although the compressed air used to operate the bell was provided by one of the two 5 hp Hornsby engines at South Stack lighthouse.

The elegant lighthouse tower at South Stack was built by Joseph Nelson to Daniel Alexander's design in 1808 and when Robert Stevenson visited ten years later, he noted that the lantern had a revolving framework containing Argand lamps and reflectors. The lantern now houses a symmetrical six panel first order Chance optic which floats in a mercury bath and completes one clockwise rotation every minute.

Access to the small island is across the suspension bridge near the foot of 300-ft high cliffs and popular stories suggest that the steps leading down the cliffs were cut by a keeper, one for each day of the year. There are over 400 concrete steps winding down the massive folds in the rock to the bridge. The cliffs are the exposed buckled strata of rocks thrusting from deep water straight up to the summit of Holyhead mountain, around 700 ft high. The rock ledges provide accommodation for thousands of seabirds, including great black backed gulls, herring gulls, razorbills, guillemots and fulmars, all raucously screeching and producing a sound which reverberates around the steep cliffs. The sound was traditionally used by local boatmen in poor visibility and I am sure some would say it was possible to get a position fix by identifying different cliff faces from the predominant colonies of birds on each. According to one source

> *At South Stack ... tamed seabirds are made use of as signals. The gulls perch on the lighthouse walls and utter loud cries which wave off approaching seamen. This*

South Stack at twilight.

*lighthouse possesses a bell and a cannon, but the natural signal has been esteemed
so superior that the cannon has been removed to a distance from the rock lest its
discharge should alarm the birds'.*[17]

Although this description has been thoughtfully embroidered, the birds certainly
are noisy! Since automation in 1984, South Stack has been managed by great black-
backed gulls and they don't care much for visitors![18]

A powerful tidal race runs off the north of South Stack, dangerous in good
visibility, let alone in fog. A large fog bell of 42 cwt was installed in 1855 which had
a 'continuous slow stroke' and was operated by clockwork. The weights probably
ran over the cliff edge, as at Start Point, and a similar bell house was constructed at
a total cost of £722 16s 6d.

The bell house no longer exists and the later horn house was demolished when
the triple-frequency fog signal was built. Other buildings have been and gone but
those remaining include the engine room and keepers' accommodation in a long
single storey adjacent to the lighthouse tower, the old oil store, coal store and wash
house near the cottage entrance.

[17]   Davenport Adams, W.H. *Lighthouses and Lightships*, 1870.
[18]   Public access has been restored to the island since my visit.

Sacks of coal and barrels of oil 'came aboard' via the winch on the landing stage not far from the bridge on the easternmost point of the island. Although the boat was afforded some shelter, a heavy swell surging between the island and the foot of the cliffs would have made the job difficult, if not dangerous. Fortunately for the keepers, they normally came across to the island on the bridge.[19]

Some steps behind the coal store lead down along a wall to the northernmost edge of the island, where there is an incline on which a unique auxiliary light was once shown in foggy weather. A small clockwork lantern was lowered on rails and the 1841 Admiralty List indicated that 'When foggy an occasional Revolving Light is shown, about 40 feet above the sea, and about 30 yards north of the High Light'. The auxiliary low light was used until about 1900.

The Trinity House Fog Signal Sub Committee proposed to install a manual reed horn at South Stack 'in lieu of the bell' in 1893, similar to those installed to replace gongs aboard lightvessels. However, a more powerful reed signal was subsequently decided upon which was installed in a new fog signal building in 1895. The reed was modified in 1909, but remained in service until October 1937, when a temporary siren was established whilst the old equipment, including the two 5 hp Hornsbys, was removed.

A twin diaphone turret was constructed on the modified horn house building and the diaphone came into service in the following March, although this installation had been originally planned in 1923.

The reed had sounded a seven second blast every 30 seconds but the character of the twin K type diaphone was a 2½ second blast every 45 seconds, requiring over 200 cubic feet of air. The compressed air was provided by single cylinder Blackstone spring injection engines coupled to Reavell Quadruplex compressors.

Blackstone and Hornsby engines were relatively quiet machines and most of the engine room noise came from the compressor, especially on starting when the unloading valves were open. Later diesel engines could be much noisier. At the nearby Skerries lighthouse, the diaphone was powered by T-type Gardners from the 1930s, as used at Lundy North and elsewhere, but when the lighthouse was refurbished in the mid-1960s, six-cylinder air-cooled Ruston engines were installed with upright Reavell compressors, an arrangement which made for a very noisy engine room. Similar equipment was installed at Pendeen in 1963.

Running a compressed air fog signal was an expensive procedure in terms of machinery, fuel and manpower but for many years compressed air signals were the most powerful instruments available and compressed air provided the most practical means of storing energy until required.

---

[19] During construction of the lighthouse, an aerial ropeway was used to transport materials to the island from the top of the cliffs. The first suspension bridge was built soon after. See Hague, Douglas B. *Lighthouses of Wales*, 1994.

As this book has shown, the types of fog signals used at lighthouses developed from the early use of guns and percussion devices to a widespread use of reeds, sirens and then diaphones, which were all operated by compressed air.

The demise of the compressed air signal was a consequence of several contributory factors, including the improvement of low power electric sound signals, the development of automatic lighthouse apparatus and the development of increasingly sophisticated onboard navigation systems.

In the early 1900s, the inventors of the submarine signalling system were convinced that sound signals in air would be redundant before 1915, but by 1929, delegates at the International Lighthouse Conference observed that the '... submarine bell is an out of date and relatively inefficient fog signal'.

The submarine signalling system introduced onboard receiving equipment which became increasingly necessary with diverse radio beacon and other direction finding systems. The radio beacon developments were ultimately succeeded by radar and racon, the long range radio networks and global positioning by satellite (GPS). The lighthouse service marked these developments with innovations of its own and an inevitable acceleration of the lighthouse automation programme was led by computer communication. Today, the lighthouse authorities have retained a few low power, high frequency electric emitters, although these are generally for the benefit of small craft.[20] It is perhaps a little surprising that new developments in satellite navigation aids and electronic charts still leave room for a nineteenth century aid to navigation like the sound signal!

From its introduction, the sound signal was primarily only a warning – it never satisfied the ambition of being an equivalent of the lighthouse for navigational purposes.

The triple-frequency fog signal at South Stack was discontinued in 1995 when Holyhead Depot was closed and the lighthouses monitored from there were transferred to the Operations Control Centre at Harwich.

Some years before the automation of South Stack, I was reliably informed by a keeper that an unusual illuminant was in use there – inside a special tube which ran down the centre of the tower was an extremely big candle! At other lighthouses, this tube contained the weights which descended inside as the clockwork rotated the optic. The weights had to be wound up at regular intervals or the clockwork stopped and the optic would cease to rotate. If the weights fell too low a warning bell rang, causing the unsuspecting keeper to leap from repose into startled activity and rush up to the lantern room to crank several dozen revolutions of the handle. It

---

[20]   It is unlikely that large vessels would be able to make much use of short range signals. It must be hoped that competent mariners and seaworthy ships can triumph over busy shipping lanes, deadlines, outflagging and equipment failure. These issues, however, are not the responsibility of the lighthouse authorities.

wasn't the task itself, but the bell which was irksome and an improved early warning consisted of a piece of folded paper neatly inserted into the weight tube to provide a discreet semaphore signal!

Many other tales and anecdotes (not all suitable for publication!) re-animated and repopulated the empty buildings I visited.

The aim of automating certain lighthouses was feasible by the start of the twentieth century but the early lighthouse designers and engineers could not have foreseen their buildings becoming unoccupied and designed them to be lived in by the people who operated and maintained the equipment.

A briny grease now builds up on the hundreds of square feet of lantern panes at an unattended lighthouse – service regulations once insisted that these were to be cleaned every day. It is not dirty inside the typical lantern room, but nothing gleams like it used to. In the past, the lantern room quickly became dirty with soot from the oil lamps and so regular cleaning was essential. The daily routine of stripping down and cleaning the PVB[21] was a chore and it probably seemed to the keepers that all the other equipment produced dirt of some kind. The fog signal engines made the floor oily and the spinning of the sirens could spray oil in a wide radius. The machinery required more attention than just Brasso and an oily rag, however, and between the engineers' visits, keepers were expected to attend to basic running repairs.

A wide range of other odd jobs also required the attention of the keepers, including the painting of the buildings and boundary walls every year, although this job was usually contracted out by the 1930s.

Prior to qualification, an Assistant Keeper had to pass certificates in the proficient operation and management of a range of equipment likely to be encountered at various lighthouses, which included competence in plumbing and carpentry besides other general duties. Management of the explosive fog signal was included in the requirements of Certificate No. 1 and No. 3 covered the working of the fog signal oil engines, compressors and sirens.

A daily fog signal allowance used to be paid to lighthouse keepers, 3d for a Principal Keeper and 2d per day for an Assistant Keeper. In the 1920s, this was increased by 2d per hour at lighthouses with a siren or a reed signal and by 1d per hour at lighthouses with fog bells. The payments were calculated from the charts used in the fog signal recorder, which indicated when the fog signal was started and how long the signal had been in operation.[22]

---

[21]   The pressurised vapour burner (PVB) was used from around WWI to the 1970s. Pressurised paraffin vapour burned brightly over a silk mantle several inches in diameter.

[22]   When the recording clocks were withdrawn in the 1960s, the hourly fog signal allowance was removed and an average amount added to keepers' pay.

Climbing the steep steps into the lantern room and seeing the beautifully constructed optic is always impressive. By comparison, the modest fog signal machinery has usually maintained a much lower profile. The way in which the glass prisms gleam and produce heavenly colours as the light refracts at their edges is dazzling and sublime; whereas, over here we have some iron pipes, trumpets and old engines …

The author of a nineteenth century Notice to Mariners could not constrain the poetic in a description of the apparent intensity of the lighthouse seen from a certain distance as '… resembling a star of the first magnitude'.[23] To help the mariner identify a fog signal, a comparable allusion in the List of Lights could only prosaically state that the explosive signal would sound a '… report similar to that of a gun'!

The sound of the diaphone under test at Chance Brothers lighthouse works was compared by a local journalist to the roar of 'a 1000 elephants', but a more poetic, if not vitriolic, description of a New England fog signal compared it to a

> '… screech like an army of panthers … gradually lowering in note until after half a minute it becomes the roar of a thousand mad bulls, with intermediate voices suggestive of the wail of a lost soul, the moan of the bottomless pit and the groan of the disabled elevator'.[24]

They don't make them like they used to! I'm not sure they ever made them like this – half a minute of a 'panther army' *before* mad bulls and elevators seems like a long time for even the most impressive fog signal to sound! To avoid the need for impossibly large quantities of stored air and impossibly large numbers of engines and compressors, most signals gave one or two blasts not exceeding ten seconds duration within a sixty second period.[25]

During the South Foreland trials in 1874, the lightshipmen on the Varne heard a noise which was '… very gross, resembling the bellowing of a bull', and from this description Professor Tyndall realised that they had heard the large American steam whistle. I never heard a bellowing bull, nor a 'thousand mad bulls', but at Lands End, the distant twelve-inch siren at Pendeen used to sound like a cow mooing.

As far as I can tell from the description, the wailing-lost-soul and bottomless-pit-moan can only be a description of the wave-actuated buoy whistle! This is a truly dismal sound. First used in the USA in 1876, the buoy uses its movement in the water to open and close a valve to trap air which it then releases through the whistle.

---

[23] From Notice to Mariners issued by the Commissioners of Northern Lighthouses, 1st February 1811, describing the Bell Rock lighthouse. To the mariner familiar with navigation by stars, this description is not merely poetic, but of genuine usefulness.

[24] Quotation from a 1905 New York newspaper in Jim Gibbs' *Lighthouses of the Pacific*, 1986.

[25] In the early days the signal might sound only once every four or five minutes. The frequency and maximum duration of the character was determined by the ability of the compressor to replenish the sounding receiver, otherwise the power of the blast would continually diminish to the point where the equipment could not produce any sound.

The moaning sound trails away and then gasps and wheezes as the lost soul goes under once more ...

Perhaps the 'mournful' description of the sound of a fog signal is attributable to its sounding in the minor key, or perhaps it's merely an associative connection with all other dull, damp, miserable foggy days! When the siren was first heard in Britain during the South Foreland trials, James Douglass learned from a local landlady that '... the fog horns ... made her quite miserable'!

Whether or not it made them miserable, lighthouse keepers had no choice but to get used to the sound of the fog signal. At some lighthouses the keepers' families also lived near the fog signal. From her new home, a keeper's wife wrote to a friend '... that horrible, mournful fog signal is still going. Will the children ever get to sleep while it is sounding?'[26] I am pleased to say I was assured by keepers that their children did get used to the noise and could sleep through it without difficulty, sometimes waking up when the fog signal was stopped! The regular rhythm of the signal could be quite comforting, although not if the sound level inside the keepers' quarters was excessive.[27] Depending on the distance from the signal, the noise level in some quarters could be loud enough to disturb sleep: where a normal bedroom might register a reading of 35 decibels, the keeper's room could be 60 dBA. At the signal itself, the sound pressure level would typically be 128 dBA. In the engine room, the reading might be between 90 and 100 dBA.[28]

The introduction of health and safety regulations in the 1970s changed responsibilities and awareness of the problems of noise. It was recognised that prolonged exposure to high noise levels could damage hearing and exposure to the short blasts of a fog signal could have a cumulative effect.

For many years, lightshipmen had been particularly vulnerable to high noise levels, but even with hindsight it is difficult to see what alternatives there were to the manned lightvessels. The fog signal was undoubtedly important[29] and the fog signal machinery demanded manual operation. If it had been possible to automate such signals at early stages, this was the place to start. In close proximity to the signal, nauseating pain could be experienced, but there was an insensibility to the

[26] Phillips, G.W. *Lighthouses and Lightships and the Men Who Man the Trinity House Lighthouse Service*, 1949.

[27] At many of the Scottish lights the siren was a little way off from the keepers' accommodation (sometimes a long way off) so that the noise levels endured were reduced by comparison with, for example, Pendeen and Hartland Point, where the accommodation was close to the signal.

[28] The threshold of hearing is 0dB sound pressure level and 3dB increments represent a doubling of the previous SPL. From 0dB to 90dB is equal to a ratio of 1,000,000,000:1. (One thousand million to one). The following shows comparative levels: whisper, 20 dB; bedroom 35, dB; quiet street, 50 dB; noisy office, 70 dB; heavy traffic, 80 dB; rivetting machine, 100 dB; high speed train, 110 dB; noise levels above 120 dB cause pain. The A-weighted noise measurements make adjustments for the difference between continuous and intermittent sound pressure levels.

[29] Besides providing a danger warning to mariners, the fog signal additionally helped to protect the lightvessel itself, which was even more vulnerable to collision in fog. The lightvessels tended to start and stop their signals with caution in mind especially since the visibility was not easily ascertained out at sea.

danger of prolonged exposure to lower levels of noise and keepers and lightshipmen simply acclimatised themselves to it. By all accounts, lightshipmen learned to communicate in the gaps between diaphone blasts, even continuing this behaviour ashore by interrupting their conversation according to the fog signal characteristic!

And some people who lived *near* fog signals thought that *they* were inconvenienced by the noise! In a number of cases, complaints made to the lighthouse authorities by the public resulted in attempts to modify the offending fog signal. Many complaints were due to 'nuisance noise' caused by the higher frequency electric emitters installed in the 1970s and 1980s. But these were not the first complaints – landowners had tried litigation to oppose the use of fog signals in the previous century![30] Where the local community appreciated the safety of shipping, the fog signal was accepted as a practical necessity and people grew used to it. Over time they may have even grown to like it.

For the mariner, the sound of a fog signal was not just a warning but could also be a reassurance. The American two-tone diaphone sounded very like a reassuring voice calling in the distance – the sounds could be wonderful! The electric LIE 300 produced a tranquil, haunting note which, when heard a mile or two distant within a background of wind, sea and bird noises, could be quite beautiful. In addition to their primary purpose, fog signals added a complementary dimension to the coastline, with their evocative sounds encouraging reflection and quietude. Their regular rhythm was superimposed on that of the waves on the sea and one's own breathing rhythm.

The lower frequency notes are nearly all consigned to history, but for a few years they will be survived by sounds between 500 Hz and 800 Hz.[31] The higher frequencies may not integrate quite so sympathetically with the ambient environmental sounds – but that is not their purpose!

Perhaps even these fog signals will be regarded with a sense of nostalgia once they have all gone.

---

[30] Munro, R.W. *Scottish Lighthouses*, 1979, see pp. 51, the Dirleton Estate near Fidra lighthouse, 1885.

[31] Signals still being installed include the Honeywell 800 Hz signal and the Pharos Marine ELG 500 and SA 850. The latter uses ELU 800 emitters sounding at 800 Hz.

One of the 4-unit 800 Hz emitters typical of fog signal equipment still in service.

Appendix I

Starting the fog signal

Sequence to start signal

Check fuel in header tank.

Switch on fuel in engine bed.

Check oil in engine with dipstick and in compressor sight glass.

Open valve on inlet side of cooling water pump.

Open outlet valve on cooling water return.

Prime hand plunger on compressor twice to lubricate system.

Switch off battery charger.

Check that double action screws on compressor cylinders are fully open.

Open pressure release valve on compressor.

Switch to 'on' position and press button on underside of fuel pump.

Press 'start' button and hold in position until engine fires.

Watch the rev counter, at around 700 rpm flywheel engages gear box and drive belts.

With rev counter at 1150, slowly screw down relief valve on compressor so that pressure gauge settles at 35 psi.

Tighten locking screw to hold valve position.

Open air discharge stop valve.

Open main stop valve on storage receivers.

Screw down both double acting screws on compressor to deliver full capacity of 35 psi to receivers.

Open stop valve to sounding receiver and check pressure gauge shows maximum of 30 psi (this will drop to 25 psi when sounding).

Switch on 50v electric supply.

Switch on the coders.

Siren should now run.

Check cooling water temperature and circulation to engine, normal temperature is 145–150°F. Lubricating oil pressure is 35 psi and temperature 125°F.

Cooling water to compressor should not exceed 160°F and will normally be about 80°F.

To stop the signal

Close stop valve on sounding receiver and switch off coders and 50v supply.

Close stop valve on main receiver and unscrew both double acting valve and compressor.

Switch off engine.

Close stop valve and circulating water supply and return. Close air discharge stop valve, open compressor relief valve.

Close fuel supply and switch on battery charger.

Wipe down and lubricate.

Approximate sounds of fog signals

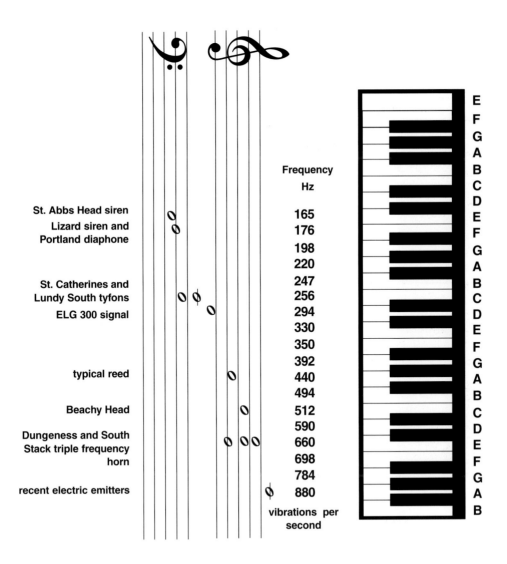

Bibliography

Abell, Sir Westcott Stile. *The Safe Sea*, Liverpool, 1932.

Adams, W.H. Davenport. *Lighthouses and Lightships*, Nelson, London, 1870.

Almqvist, E. and Sutton-Jones, K. *Milestones in Lighthouse Engineering*, Pharos Marine, 1994.

Atkinson, Frank. *Victorian Britain: The North East*, David & Charles, 1989.

Babbage, Charles. *Notes Respecting Lighthouses*, 1852.

Beaver, Patrick. *A History of Lighthouses*, Peter Davies, 1971.

Beazeley, Alexander. *On Coast Fog Signals*, MPICE, 1872.

Beazeley, Alexander. *On Phonic Coast Fog Signals* No.1296, MPICE, 1871.

Blake, George. *Clyde Lighthouses - Clyde Lighthouse Trust 1756–1956*, 1956.

Bowen, Sir John P. *British Lighthouses*, British Council, 1947.

Burton, Anthony. *The Past Afloat*, Andre Deutsch, 1982.

Case, A.H. *Beachy Head Lighthouse*, MPICE, vol. 159, 1905.

Chadwick, Lee. *Lighthouses and Lightships*, Dennis Dobson, 1971.

Chapman, C.W. *Modern High Speed Oil Engines*, vols. 1, 2 & 3, Caxton, 1949.

Cowper, Frederick. *Sailing Tours*, vols. 1–4 (reprinted, Ashford Press 1985), 1895.

Cunningham, Alexander. *Fog Signals*, Royal Scottish Society of Arts, 1863.

Davis, G.M. and R.C. *Trial of Error*, 1983.

Douglass, Sir James N. *The Wolf Rock Lighthouse* No. 1268 MPICE, 1870.

Douglass, Sir James N. *Electric Light Applied to Lighthouses*, MPICE vol. 57, 1879.

Douglass, Sir James N. *Improvements in Coast Signals*, 1884.

Edgington, D. and Hudson, C. *Stationary Engines for the Enthusiast*, 1981.

Edwards, E. Price. *A List and Description of Coast Fog Signals*, 1880.

Edwards, E. Price. *Our Seamarks*, 1884.

Elliot, Maj. G.H. *Report of a Tour of Inspection of European Lighthouse Establishments*, Washington, 1874.

Esquiros, Alphonse. *Cornwall and Its Coasts*, 1865.

Findlay, A. *Lighthouses, Fog Signals and Tides*.

Findlay, A. *Lighthouses of the World*, 1861 and 1900.

Gedye, N.J. *On Coast Fog Signals*, MPICE, 1902.

Goldsmith-Carter, G. *Looming Lights*, Readers Union, 1947.

Gibbs, J.A. *Lighthouses of the Pacific*, Schiffer, 1986.

Hague, Douglas and Christie, Rosemary. *Lighthouses, Their Architecture, History and Archaeology*, Gower Press, 1975.

Hague, Douglas (Ed. S. Hughes). *Lighthouses of Wales*, RCAHMW, 1994.

Hardy, W.J. *Lighthouses, Their History and Romance*, 1895.

Hobbs, J.S. *Bristol Channel Pilot, 1859* (Reprinted D. Bradford Barton, 1972) 1859.

Ingrey, Charles. *Ailsa Craig Fog Signalling Machinery*, MPICE, 1886.

Jackson, D. *Lighthouses of England and Wales*, 1975.

Jenkins, Henry Davenport. *Lights and Tides of the World*, Imrey, London, 1900.

Jenkins, Henry Davenport. *The Pilots Guide for the English Channel*, 1912.

Johnson, A.B. *Aberrations of Audibility of Fog Signals*, Washington, 1885.

Langham, M. *A Lundy Album*, 1980.

Langton-Jones, R. *Silent Sentinels*, London, 1944.

MacCormick, W.H. *The Modern Book of Lighthouses*, 1936.

Majdalany, F. *The Red Rocks of Eddystone*, Longmans, 1954.

Millar, G.L. *The Recent Revolution in Organ Building*, Charles Francis Press, New York, 1913.

Munro, R.W. *Scottish Lighthouses*, Thule Press, 1979.

Naish, John. *Seamarks*, Stanford Maritime, 1985.

Nicholson, C. *Rock Lighthouses of Britain – the End of an Era?*, Whittles Publishing, 1995.

Noall, Cyril. *Cornish Lights and Wrecks*, 1968.

Phillips, Godfrey W. *Lighthouses and Lightships and the Men Who Man the Trinity House Service*, Robert Ross & Co, London, 1949.

Plimsoll, Samuel. *Our Seamen: An Appeal*, 1873.

Rees, John S. *History of the Liverpool Pilotage Service*, Southport, 1949.

Rendell, Joan. *Lundy Island*, Bossiney Books, 1979.

Richardson, E.G. *Sound*, 1929.

Singer, C. (Ed.) *A History of Technology*, Oxford, 1954.

Stevenson, David A. *On Coast Fog Signals*, Edinburgh, 1881.

Stevenson, David A. *Ailsa Craig Lighthouse and Fog Signal*, MPICE, vol. 89, 1886.

Stevenson, D. Alan (Ed.) *English Lighthouse Tours, 1801, 1811, 1818*, 1946.

Stewart-Eve, A. and Creasey, C.H. *The Life and Work of John Tyndall*, Macmillan, London, 1945.

Stone, John Harris. *England's Riviera*, c.1910.

Strutt, Robert J. *The Life of John William Strutt, 3rd Baron Rayleigh, O.M., F.R.S.*, (revised edition, 1968), 1924.

Sutton-Jones, K. *Pharos – The Lighthouse Yesterday, Today and Tomorrow*, Michael Russell, 1985.

Talbot, F.A. *Lightships and Lighthouses*, Heinemann, London, 1913.

Tyndall, John. *Sound*, 1867.

Tyndall, John. *Recent Experiments in Fog Signals*, Proc. of Royal Society. 1878.

Williams, Thomas. *The Life of Sir James Nicholas Douglass*, Longmans, 1900.

Woodman, Richard. *Keepers of the Sea*, Terence Dalton, 1983.

Woodman, Richard. *View From the Sea*, Century, London, 1985.

(Note: MPICE – Minutes of the Proceedings of the Institution of Civil Engineers)

Articles, documents and journals

Admiralty Hydrographic Office Charts.

Admiralty List of Lights, from 1850.

Admiralty *Manual of Navigation*.

A Few Notes on Modern Lighthouse Practice, Chance Bros. Ltd., 1910.

A.G.A. Brochures and Information Leaflets.

A Report by Professor Tyndall to the Trinity House upon Recent Experiments with regard to Fog Signals, 1874.

Clingan, I.C. 'Electrical Equipment in Lighthouses' in *Electrical Industries Export*, March, 1960.

Fox, H. 'The Lizard Lighthouses' in *Journal of the Royal Institute of Cornwall*, vol. 6. 1881.

London International Lighthouse Conference, 1929.

Parliamentary Proceedings, 1870–1882 (Lighthouse Illuminants).

Report of Trinity House Fog Signal Committee on Experiments conducted at St Catherine's Point, Isle of Wight, 1901.

Sayer, M. 'New Light on Hope Jones' in *The Organ* LX (1980–1981).

Trinity House Report on the Proceedings of the International Technical Conference on Lighthouses and Other Aids to Navigation (Paris 1933), HMSO, 1936.

Trinity House Notices to Mariners.

Electrical Engineering, vol. 29. 1957.

Engineering, from 1864.

Flash, Trinity House Magazine

LAMP, The Newsletter of the Association of Lighthouse Keepers.

The Engineer

The Syren, 1896–1897.

The Syren and *Shipping Illustrated* from 1897.

Unpublished Documents

Ailsa Craig – Instruction Manual for Keepers, 1886.

Instructions to Light Keepers, US Lighthouse Board, Washington, 1881–1902.

Lighthouse and Lightvessel Order Books, Trinity House.

Manual of Fog Signal Operating Instructions, Trinity House Archives, Guildhall Library.

Minutes of The Proceedings of The Fog Signal Subcommittee, 1892–1893, Guildhall Library.

Miscellaneous correspondence and reports relating to fog signals, 1874–1878. Trinity House.

Northern Lighthouse Board Engineering Drawings, National Monument Record of Scotland.

Reports of the Scientific Advisor to Trinity House.

Steven and Struthers Records, University of Glasgow.

Trinity House Engineering Drawings.

Index

Lighthouses and lightvessels (LTVS) shown in upper case, italics indicate non-Trinity House/ NLB lights.

Page numbers in bold italics indicate an illustration in the text.